Chemical Graph Theory

Volume I

Author
Nenad Trinajstić
Professor of Chemistry
The Rugjer Bošković Institute
Zagreb, Croatia
Yugoslavia

CRC Press, Inc.
Boca Raton, Florida

Library of Congress Cataloging in Publication Data
Trinajstic, Nenad, 1936-
Chemical graph theory.
Includes bibliographical references and index.
1. Chemistry—Mathematics. 2. Graph theory.
I. Title.
QD39.3.M3T74 1983 541.2'2'015115 82-12895
ISBN 0-8493-5273-8 (v.1)
ISBN 0-8493-5274-6 (v.2)

Direct all inquiries to CRC Press, Inc., 2000 Corporate Blvd., N.W., Boca Raton, Florida, 33431.

International Standard Book Number 0-8493-5273-8 (Volume I)
International Standard Book Number 0-8493-5274-6 (Volume II)
Library of Congress Card Number 82-12895
Printed in the United States

To my brother Ivan

PREFACE

The greater part of this book was written during my visit to the Department of Chemistry, the University of South Carolina at Columbia in late autumn 1979 and winter 1980. I felt for some time that an introductory book on chemical graph theory is needed, because it is considered by many as a new branch of chemistry, though its beginnings go back to over 100 years ago. Since I was asked to deliver a series of lectures on chemical graph theory, I used the notes for these lectures as an embryo from which the present book has evolved.

This book is entitled *Chemical Graph Theory* because we will be mostly concerned with the handling of the chemical (molecular) graphs, i.e., mathematical diagrams representing molecular structures. Thus, chemical graph theory is engaged with analyses of all consequences of a connectivity in a system, e.g., bonds in molecules or reaction paths in chemical transformations. Ordinarily, graph theory does not produce numerical data (just as group theory), it uses available data and searches for regularities that can be attributed to a combinatorial and topological origin. Therefore, here is a good opportunity to indicate clearly for the chemical community at large, in order to avoid possible misunderstandings of the use and potential of (chemical) graph theory, that graph theoretical methods and proof techniques should be expected to be used as a complementary approach where the topology and the combinatorial nature play an important role, in parallel to the application of the group theory to problems where symmetry is an important characteristic of the system. Besides in the chemical graph theory we are allowed to rely on the intuitive understanding of many concepts and theorems rather than on formal mathematical proofs.

The roots of chemical graph theory may be found in the work by Higgins,[1] who has used first chemical graphs (albeit not recognized as such) for representing molecules; Kopp,[2] who studied the additive properties of molecules; and Crum Brown,[3] who has made use of quite a modern graphic notation for chemical compounds. Other early contributors to chemical graph theory are Laurent,[4] who discovered that the number of atoms with odd valency in a molecule is always even; Cayley,[5-7] who developed the mathematical theory of isomers (alkanes) using graphs called trees; Flavitzky,[8] who calculated the number of isomers of saturated alcohols; and Sylvester,[9] who first pointed out the analogy between chemistry and algebra. Sylvester's work is especially important to us because he clearly expressed a belief that there is a common ground for the interchange of mathematical and chemical ideas which may lead to new development of both fields what was the case in many instances since his time.[10-15]

This book will be mainly concerned with the structural aspects of chemical graph theory. It may be partitioned in three parts: (1) the elements of graph theory needed for the understanding of what follows; (2) the topological aspects of Hückel theory, resonance theory, and theories of aromaticity; and (3) the applications of chemical graph theory to the structure-property and structure-activity relationships and to the isomer enumeration. Each chapter is followed by a list of the pertinent references, where the additional material about the work referred to may be found.

I am grateful to many people who directly or indirectly influenced me to undertake this project. Professor Benjamin M. Gimarc (Columbia, S.C.) convinced me that I should write this book in the light of the role of the Zagreb Group in the remarkable growth of the chemical applications of graph theory in the last decade. He was kind enough to read the entire book in the manuscript form and to give useful comments leading to the improvement of the book in every way. This is a good opportunity to thank him for critical, but friendly discussions, comments, and help.

P. Křivka (Pardubice), I. Mladenov (Sofia), and B. Mohar (Ljubljana) also commented on several chapters of the book. I am thankful to them for their kind help.

I wish to thank several people for helping me to enter the fascinating field of (chemical) graph theory. I am indebted to Professor A. T. Balaban (Bucharest), Professor F. Harary (Ann Arbor), and Professor A. J. Schwenk (Annapolis) for teaching me the elements of graph theory. I am also thankful to Professor W. C. Herndon (El Paso), Professor B. A. Hess, Jr. (Nashville), Professor H. Hosoya (Tokyo), Professor L. Klasinc (Zagreb), Dr. R. B. Mallion (Canterbury), Professor O. E. Polansky (Mülheim/Ruhr), Professor V. Prelog (Zürich), Professor M. Randić (Ames), and Professor L. J. Schaad (Nashville) for encouragement and support over the years.

Finally, I wish to thank Dr. B. Džonova-Jerman-Blažić (Ljubljana), Dr. I. Gutman (Kragujevac), Dr. M. Milun (Zagreb), and Dr. P. Ilic (Sarajevo), my former graduate students, for their enthusiasm and hard work, and the past and present members, and the guests, of the Zagreb Group: Dr. S. Bosamac, Dr. D. Bonchev (Burgas), Mrs. M. Barysz (Katowice), Dr. A. Graovac, Mr. Ž. Jeričević, Mr. A. Jurić (Banja Luka), Mr. D. Kasum, Professor J. V. Knop (Düsseldorf), Dr. A. Sabljić, Dr. J. Seibert (Hradec Králové), Dr. A. Velenik-Oliva, Professor C. F. Wilcox, Jr. (Ithaca), and Dr. T. Živković for their patient collaboration during the past 14 years.

REFERENCES

1. **Higgins, W.,** *A Comparative View of the Phlogistic and Anti-Phlogistic Theories,* Murray, London, 1789.
2. **Kopp, H.,** *Ann. Chem.,* 41, 79, 1842; 41, 169, 1842.
3. **Crum Brown, A.,** *Trans. R. Soc. Edinburgh,* 23, 707, 1864.
4. **Laurent, A.,** *Ann. Chim. Phys.,* 18, 266, 1864.
5. **Cayley, A.,** *Phil. Mag.,* 18, 374, 1859.
6. **Cayley, A.,** *Phil. Mag.,* 47, 444, 1874.
7. **Cayley, A.,** *Chem. Ber.,* 8, 1056, 1875.
8. **Flavitzky, F.,** *J. Russ. Chem. Soc.,* 160, 1871.
9. **Sylvester, J. J.,** *Nature (London),* 17, 284, 1877—1878; *Am. J. Math.,* 1, 64, 1878.
10. **Biggs, N. L., Lloyd, E. K., and Wilson, R. J.,** *Graph Theory 1736—1936,* Clarendon, Oxford, 1976.
11. **Rouvray, D. H.,** *R. I. C. Rev.,* 4, 173, 1971.
12. **Mallion, R. B.,** *Chem. Br.,* 9, 242, 1973.
13. **Gutman, I. and Trinajstić, N.,** *Topics Curr. Chem.,* 42, 49, 1973.
14. **Wilson, R. J.,** in *Colloquia Mathematica Societatis János Bolyai,* Vol. 18 Combinatorics, Kezthely (Hungary), 1976, 1147.
15. **Cvetković, D., Doob, M., and Sachs, H.,** *Spectra of Graphs,* Academic Press, New York, 1980.

THE AUTHOR

Nenad Trinajstić is a Senior Researcher at the Rugjer Bošković Institute in Zagreb, Croatia, Yugoslavia and a Professor at the Department of Chemistry, Faculty of Science and Mathematics, University of Zagreb. He received the B.Sc. (1960), M.Sc. (1966), and Ph.D. (1967) degrees from the University of Zagreb. During the period 1964 to 1966, he was a predoctoral fellow with Prof. John N. Murrell at the University of Sheffield and the University of Sussex, U.K. and collaborated with him in research on MO interpretation of electronic spectra of conjugated molecules and on the development of criteria for producing localized orbitals. Dr. Trinajstić's postdoctoral years (1968 to 1970) were spent with Prof. Michael J. S. Dewar at the University of Texas at Austin. The main research interest during this period was the development of a convenient semiempirical MO theory for studying the ground states of large molecules. After returning to Zagreb from the U.S., he initiated research on chemical applications of graph theory and has greatly contributed to the revival of the uses of graph theory in chemistry. In the last decade, he spent some time at various universities such as the University of Trieste, Italy (collaborating with Prof. Vinicio Galasso), the University of Utah, Salt Lake City (collaborating with Prof. Frank E. Harris), the University of Oxford, U.K. (collaborating with Dr. Roger B. Mallion), the University of South Carolina, Columbia (collaborating with Prof. Benjamin M. Gimarc), the University of Düsseldorf, West Germany (collaborating with Prof. Jan V. Knop), the Iowa State University, Ames (collaborating with Prof. Milan Randić), Higher School of Chemical Technology, Burgas, Bulgaria (collaborating with Dr. Danail Bonchev), and the University of Sussex, Brighton, England (collaborating with Prof. John N. Murrell). He had a number of B.Sc., M.Sc., and Ph.D. students, and has published over 200 papers in the fields of Theoretical Organic Chemistry and Mathematical Chemistry. His present main research interest is the chemical applications of graph theory. In 1972, he received the City of Zagreb Award for research. In 1982, he received The Rugjer Bošković Award for his work in chemical graph theory. He is married (Judita), has two children (daughter: Regina and son: Dean), and a grandson (Sebastijan).

TABLE OF CONTENTS

Volume I

Chapter 8

TABLE OF CONTENTS

Volume II

Chapter 1

INTRODUCTION

Graph theory is a branch of mathematics that deals with the way objects are connected.[1] Thus, the connectivity in a system is a fundamental quality of graph theory. Graph theory is related to matrix theory, group theory, set theory, probability, combinatorics, numerical analysis, and topology. It has been used in such diverse fields as economics[2] and theoretical physics,[3] psychology[4] and nuclear physics,[5] biomathematics[6] and linguistics,[7] sociology[8] and zoology,[9] technology[10] and anthropology,[11] computer science[12] and geography,[13] etc. Similarly, the recent years have witnessed a remarkable growth in the applications of graph theoretical principles to chemistry.[14-21] In fact, graph theory serves as a mathematical model for any abstract or real chemical system involving a binary relation.

There are several reasons for the increasing popularity of graph theory in chemistry. First, there is hardly any concept in the natural sciences which is closer to the notion of graph than the structural (constitutional) formula of a chemical compound,[22] because a graph is, simply said, a mathematical structure which may be used directly to represent a molecule when the only property considered is the internal connectivity, i.e., whether or not a chemical bond joins two atoms in a molecule. Here, the chemical bond is represented by only a line connecting two atoms. At this qualitative level, we retain the image of the simple bond which is lost in the beautiful density diagrams of today.[23] It appears that there are many areas of chemistry in which the molecular connectivity determines the properties.[14-21] Second, graph theory provides simple rules by which experimental chemists may obtain many useful qualitative predictions about the structure and reactivity of various compounds. All these predictions can be reached using nothing more than pencil and paper. Furthermore, the obtained results have a general validity and may be formulated as theorems and/or rules which can then be applied to a variety of similar problems without any further numerical or conceptual work. Third, graph theory may be used as a foundation for the representation and categorization of a very large number of chemical systems.[24,25] Moreover, chemists know and use a number of graph theoretical theorems without being aware of this fact in many cases. A classic example is provided by the concept of alternant hydrocarbons, used in chemistry[26] since 1940, which is for graph theorists the two-color problem.[1] Therefore, chemists can easily grasp the basis of graph theory. However, the language of graph theory is different from that of chemistry. Since there is no unique graph theoretical terminology,[1,27,28] we offer a short glossary in Table 1 which contains the terminology of graph theory which we propose for standard use in chemistry and the corresponding chemical terms.

In relating graph theoretical terms and chemical terms, a certain caution is needed, because one set of terms is taken from the abstract theory and the other from the concrete models used in chemistry. Thus, in one case, for example, trees may be used to represent acyclic structures and in another, certain reaction schemes.

Table 1
THE CORRESPONDENCE BETWEEN THE GRAPH THEORETICAL AND CHEMICAL TERMS

Graph theoretical term	Chemical term
Chemical (molecular) graph	Structural formula
Vertex	Atom
Weighted vertex	Atom of a specified element
Edge	Chemical bond
Weighted edge	Chemical bond between the specified elements
Degree (valency) of a vertex	Valency of an atom
Tree graph	Acyclic structure
Chain	Linear polyene
Cycle	Annulene
Bipartite (bichromatic) graph	Alternant chemical structure
Nonbipartite graph	Nonalternant chemical structure
Kekulé graph (1-factor)	Kekulé structure
Adjacency matrix **A**	Hückel (topological) matrix
Characteristic polynomial	Secular polynomial
Eigenvalue of **A**	Eigenvalue of Hückel matrix
Eigenvector of **A**	Hückel (topological) molecular orbital
Positive eigenvalue	Bonding energy level
Zero eigenvalue	Nonbonding energy level
Negative eigenvalue	Antibonding energy level
Graph spectral theory	Hückel theory

REFERENCES

1. **Wilson, R. J.,** *Introduction to Graph Theory,* Oliver and Boyd, Edinburgh, 1972.
2. **Avondo-Bodino, G.,** *Economic Applications of the Theory of Graphs,* Gordon & Breach, New York, 1962.
3. **Harary, F., Ed.,** *Graph Theory and Theoretical Physics,* Academic Press, New York, 1967.
4. **Cartwright, D. and Harary, F.,** *Psychol. Rev.,* 63, 277, 1963.
5. **Mattuck, R. D.,** *A guide to Feynman Diagrams in the Many-Body Problem,* McGraw-Hill, New York, 1967.
6. **Lane, R.,** *Elemente der Graphentheorie und ihre Anwendung in den biologischen Wissenschaften,* Akademischer Verlag, Leipzig, 1970.
7. **Čulik, K.,** *Application of Graph Theory to Mathematical Logics and Linguistics,* Czechoslovak Academy of Sciences, Prague, 1964.
8. **Flament, C.,** *Applications of Graph Theory to Group Structure,* Prentice-Hall, Englewood Cliffs, N.J., 1963.
9. **Lissowski, A.,** *Acta Protozool.,* 11, 131, 1971.
10. **Korach, M. and Haskó, L.,** *Acta Chim. Acad. Sci. Hungaricae,* 72, 77, 1972; see also Kémiai technológiai rendszerek gráfelmeléti viszgálata, Akademiai Kiadó, Budapest, 1975.
11. **Hage, P.,** *J. Polynesian Soc.,* 86, 27, 1977; Ann. Rev. Anthropol., 8, 115, 1979.
12. **Even, S.,** *Graph Algorithms,* Pitman, London, 1979.
13. **Cliff, A., Haggett, P., and Ord, K.,** in *Applications of Graph Theory,* Wilson, R. J. and Beineke, L. W., Eds., Academic Press, London, 1979, 293.
14. **Rouvray, D. H.,** *R.I.C. Rev.,* 4, 173, 1971.
15. **Gutman, I. and Trinajstić, N.,** *Topics Curr. Chem.,* 42, 49, 1973.
16. **Balaban, A. T., Ed.,** *Chemical Applications of Graph Theory,* Academic Press, London, 1976.
17. **Wilson, R. J.,** in *Colloquia Mathematica Societatis János Bolyai,* Vol. 18, Combinatorics, Keszthely (Hungary), 1976, 1147.
18. **Kier, L. B. and Hall, L. H.,** *Molecular Connectivity in Chemistry and Drug Research,* Academic Press, New York, 1976.
19. **Graovac, A., Gutman, I., and Trinajstić, N.,** *Topological Approach to the Chemistry of Conjugated Molecules, Lectures Notes in Chemistry,* Vol. 4, Springer, Berlin, 1977.

20. **Slanina, Z.,** *Chem. Listy,* 72, 1, 1978.
21. **Rouvray, D. H. and Balaban, A. T.,** in *Applications of Graph Theory,* Wilson, R. J. and Beineke, L. W., Eds., Academic Press, London, 1979, 177.
22. **Prelog, V.,** Nobel Lecture, December 12, 1975; reprinted in *Science,* 193, 17, 1976.
23. **Wahl, A. C.,** *Science,* 151, 961, 1966.
24. **Lynch, M. J., Harrison, J. M., Town, V. G., and Ash, J. E.,** *Computer Handling of Chemical Structure Information,* MacDonald, London, 1971.
25. **Carthart, R. E., Smith, D. H., Brown, H., and Sridharan, N. S.,** *J. Chem. Inf. Comp. Sci.,* 15, 124, 1975.
26. **Coulson, C. A. and Rushbrooke, G. S.,** *Proc. Cambridge Phil. Soc.,* 36, 193, 1940.
27. **Essam, J. W. and Fisher, M. E.,** *Rev. Mod. Phys.,* 42, 272, 1970.
28. **Harary, F.,** *Graph Theory,* Addison-Wesley, Reading, Mass., 1971, second printing.

Chapter 2

ELEMENTS OF GRAPH THEORY

This chapter gives the basic definitions, theorems, and concepts of graph theory. Since this book is designed for the chemical community at large, mathematical rigor is omitted whenever possible. The precise details of graph theory can be found in several excellent mathematical books dealing with the subject.[1-9]

I. DEFINITION OF THE GRAPH

The Decartes product V × W of sets V and W is defined as,

$$V \times W = \{(v,w) \mid v \in V,\ w \in W\} \tag{1}$$

i.e., V × W is the set of ordered pairs, where the first member in the pair is from the set V and the second from the set W, respectively. A *binary relation* R in V × W is any subset of V × W. When V = W, R is called a binary relation defined on the set V. The meaning of this is that two elements of the set V (v_i and v_j) either belong to the binary relation (i.e., they are connected),

$$(v_i, v_j) \in R \tag{2}$$

or they do not belong to the binary relation (i.e., they are not connected),

$$(v_i, v_j) \notin R \tag{3}$$

A *graph* is symbolized by G and is defined as an ordered pair consisting of two sets V = V(G) and R = R(G),

$$G = [V(G), R(G)] \tag{4}$$

Elements of a set V(G) are called *vertices* and elements of a set R(G), involving the binary relations between the vertices, are called *edges*. The term graph was first used by Sylvester[10] in 1878.

The above abstracts definition of a graph can be visualized when the vertices are drawn as small circles and two vertices are connected by a line if they belong to the binary relation R. Mainly because of their diagrammatic representation, graphs have an intuitive appeal for chemists.

Example

$$V(G_1) = (1,2,3,4)$$

$$R(G_1) = [(1,2), (2,1), (2,3), (2,4), (3,2), (3,4), (4,2), (4,3)]$$

$$G_1 = [V(G_1), R(G_1)]$$

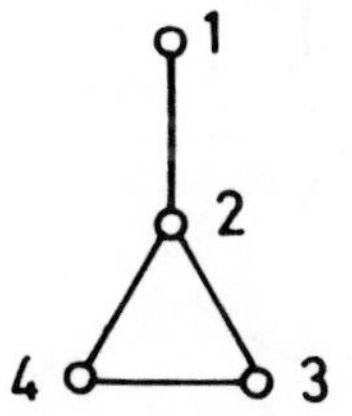

G_1

Graph G_1 is an *unoriented graph,* because it has all edges unoriented. A graph having only oriented edges is a *directed graph* or *diagraph.* An *oriented graph* is a directed graph without symmetric pairs of directed edges.

Example

G_2

G_3

G_2 and G_3 are both directed graphs, but only G_3 is an oriented graph. The number of directed graphs by far exceeds that of unoriented graphs. As an example, there are 156 unoriented graphs and 1,540,944 directed graphs with six vertices.

If two vertices are joined by more than one edge in a graph, it is called a *multigraph.* If loops, single or multiple, are allowed in a graph, it is called a *pseudograph.* A loop is an edge connecting a vertex to itself. The orientation of a loop has ordinarily no meaning and therefore is usually omitted.

Example

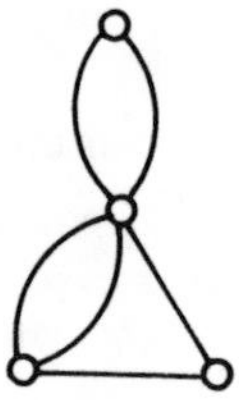

G_4

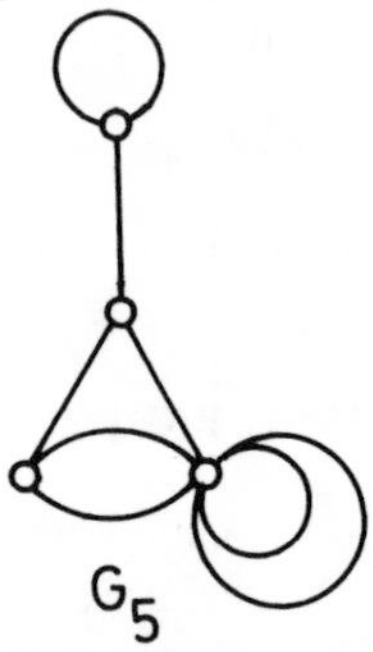

G_5

G_4 and G_5 are both multigraphs, but G_5 is also a pseudograph. The first multigraph to be used was the graph of the Königsberg bridge problem (G_6) set by Euler[11] in 1736. Euler, studying the Königsberg bridge problem, started graph theory[12] and developed a criterion for a given graph to be traversable. A graph G is traversable if it is possible to find a walk that traverses each edge once, goes through all vertices, and ends at the starting vertex.[5] (For a definition of walk, see Section III.) Thus, a traversable graph is called an *Eulerian graph.* A nontraversable graph (such as G_6), i.e., a graph for which it is not possible to go over all edges exactly once and return to the starting vertex, is named a *non-Eulerian graph.*

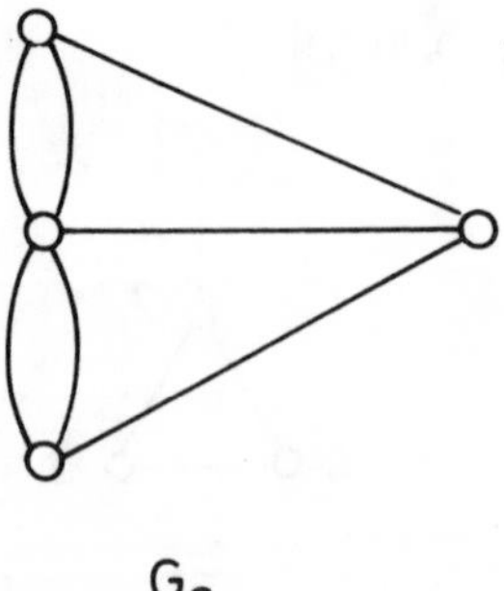

G_6

In this book, if not otherwise stated, a graph will be understood as unoriented graph without loops or multiple edges. For such a graph the binary relation R is *symmetric* and *antireflexive,*

$$(v_1,v_2) \in R \Longleftrightarrow (v_2,v_1) \in R \tag{5}$$

$$(v_1,v_2) \in R \Longrightarrow v_1 \neq v_2 \tag{6}$$

The property (5) follows from the requirement for a graph to be undirected, while the property (6) reflects the nonexistence of a loop in a graph. Properties (5) and (6) allow one to define an unoriented graph G in an alternative way. A graph G defined as,

$$G = [V(G), E(G)] \tag{7}$$

consists of nonempty set of V(G) of arbitrary labeled vertices and of a nonempty set of E(G) of unordered pairs of distinct elements of V(G) called edges. In most cases of interest for chemistry the *vertex-set* V(G) and the *edge-set* E(G) will be finite.

Example

$$V(G_7) = (1,2,3,4)$$

$$E(G_7) = [(1,2), (1,3), (1,4), (2,3), (3,4)]$$

$$e_1 = (1,2),\ e_2 = (1,3),\ e_3 = (1,4),\ e_4 = (2,3),\ e_5 = (3,4)$$

$$G_7 = [V(G_7), E(G_7)]$$

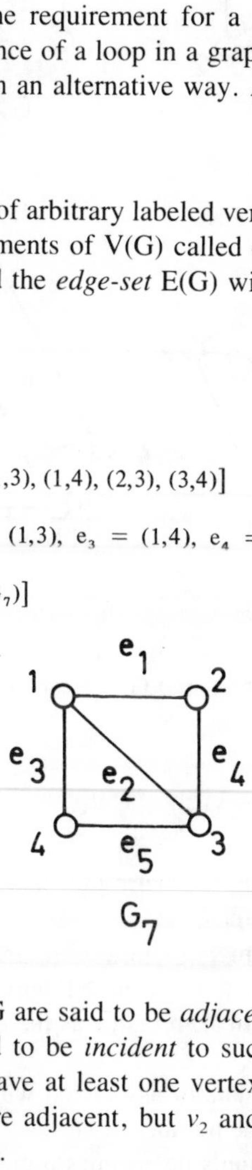

G_7

Two vertices *p* and *q* of a graph G are said to be *adjacent* if there is an edge joining them; the vertices *p* and *q* are then said to be *incident* to such an edge. Similarly, two distinct edges of G are *adjacent* if they have at least one vertex in common. In the graph G_7, for example, the vertices v_1 and v_3 are adjacent, but v_2 and v_4 are not, while edges e_1 and e_4 are adjacent, but e_1 and e_5 are not.

In mathematical literature may be found some other more general, definitions of a graph.[1,13,14]

II. ISOMORPHISM OF THE GRAPHS

Two graphs $G' = [V(G'), E(G')]$ and $G'' = [V(G''), E(G'')]$ are *isomorphic* (written $G' \cong G''$ or sometimes $G' = G''$) if there exists an one-to-one mapping *f*,

$$fv' = v'' \tag{8}$$

such that $(fv'_1, fv'_2) \in E(G'')$ if, and only if, $(v'_1, v'_2) \in E(G')$. The procedure to recognize isomorphic graphs is simple for small graphs (like G_7 and G_8 below), but it is very difficult to recognize the complex graphs (like G_9 and G_{10} below) as identical graphs.

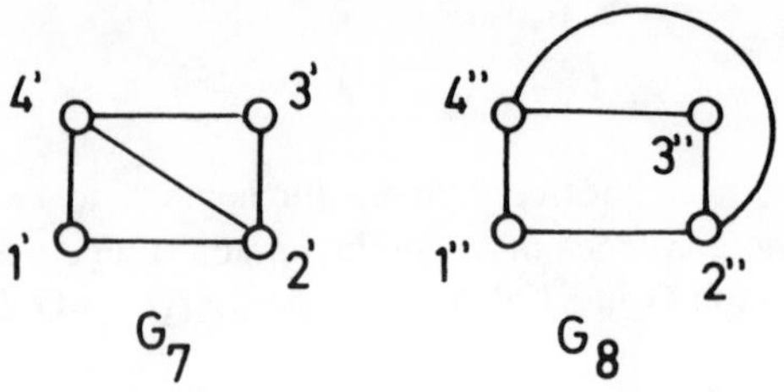

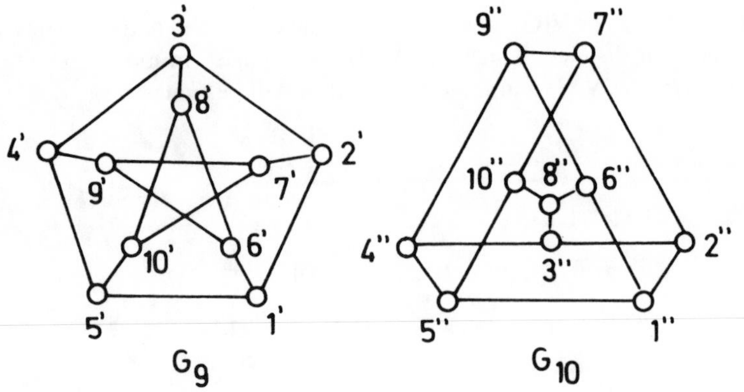

Both pairs of graphs: G_7 and G_8, and G_9 and G_{10}, are isomorphic because mapping *f* defined as

$$fv'_i = v''_i \text{ for each } i \ (v'_i \in V(G'), v''_i \in V(G'')) \tag{9}$$

preserves adjacency. It follows from the definition that the isomorphic graphs are indeed identical, but differently drawn, graphs.

Generally, the problem of recognizing isomorphic graphs is one of the grand unsolved problems of graph theory. Construction of all N! possible mappings from one graph to another, although obviously impractical, remains as the only secure check for isomorphism of graphs.

An *invariant* of a graph G is a quantity associated with G which has the same value for any graph isomorphic to G. Thus, the number of vertices and the number of edges are graph invariants. A complete set of invariants determines a graph up to isomorphism.

III. WALKS, PATHS, AND VALENCIES IN GRAPHS

A *walk* in a graph G is a sequence of edges $e_1, e_2, \ldots, e_\ell$ with the requirement that sequential edges are incident. The *length* of such a walk is ℓ. Alternatively, if a walk is represented by a sequence of vertices $v_1, v_2, \ldots, v_m$, it is called a v_1-v_m walk because it joins vertices v_1 and v_m. v_1 is called the *initial vertex* of the walk, while v_m is called the *terminal vertex* of the walk. A *closed walk* is a v_i-v_i walk, i.e., a walk which starts and ends at the same vertex v_i. Otherwise, a walk is said to be an *open walk*. If the vertices $v_1, v_2, \ldots, v_m$ are distinct, then the walk is called a *path*. If all the edges are distinct, then the walk is called *trail*.

Example

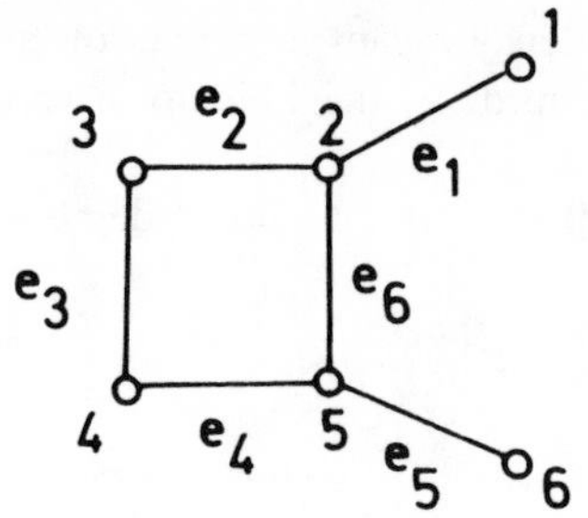

G_{11}

In the labeled graph G_{11} a sequence of vertices v_1,v_2,v_3,v_2,v_5 is a walk which may be denoted as a walk 12325. One of the paths of length four in G_{11} is, for example, the sequence $v_1v_2v_3v_4v_5$ which may be also denoted as 12345. Walks 12521 and 121 are closed, while walks 345, 123256, and 12543 are open. Walks 345 and 12543 are paths of length 2 and length 4, respectively.

The length of the shortest path between the two vertices *p* and *q* in a graph is called the *distance* between these two vertices and is denoted as *d*(p,q). The distance *d* is a nonnegative quantity and has only integral values. It has the following properties,

$$d(p,q) = 0 \text{ if, and only if, } p = q \tag{10}$$

$$d(p,q) = d(q,p) \tag{11}$$

$$d(p,r) + d(r,q) \geqslant d(p,q) \tag{12}$$

According to the definition of the graph,

$$d(p,q) = 1 \text{ if, and only if, } (p,q) \in E(G) \tag{13}$$

If every pair of vertices in G is joined by a path, G is a *connected graph*. If there is no path between two vertices in G, i.e., *d*(p,q) = ∞, G is a *disconnected graph* and these two vertices belong to different *components* of a graph.

Example

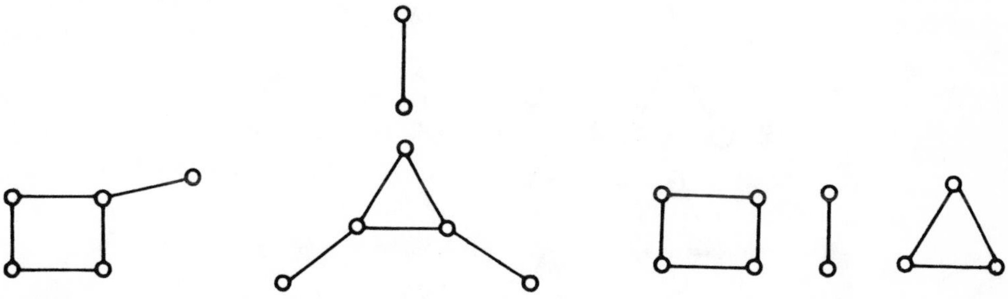

G_{12} G_{13} G_{14}

The graphs G_{12}, G_{13}, and G_{14} have one, two, and three components, respectively.

Once the distance in a graph is introduced, it is easy to define the graph theoretical *valency* or the *degree of a vertex*. All vertices among which the distance is unity (i.e., adjacent vertices) are the first neighbors. The second, the third, and the higher neighbors are defined in analogous ways. The number of the first neighbors of a vertex *p* is a degree of a vertex *p*. It is denoted as $D_1(p)$, or D(p), or simply D. D equals the number of edges incident with *p*. Thus, the graph theoretical valency means a number of connections by a given vertex. The numbers of the second, the third,..., neighbors of a vertex are consequently denoted by $D_2(p)$, $D_3(p)$,....

A *handshaking lemma* states that the total sum of degrees in a graph equals twice the number of edges,

$$\sum_{i=1}^{N} D_1(i) = 2M \tag{14}$$

where N and M stand for the total number of vertices and edges of G, respectively. The factor 2 is present in (14) because in the summation each edge is counted twice. Result (14) was known to Euler and is referred to as the handshaking lemma because it means that if several persons shake hands, the total number of hands shaken must be *even*. This is so because two hands are involved in each handshake. An immediate corollary of the handshaking lemma is that in any graph the number of vertices of *odd* degree must be *even*.

Example

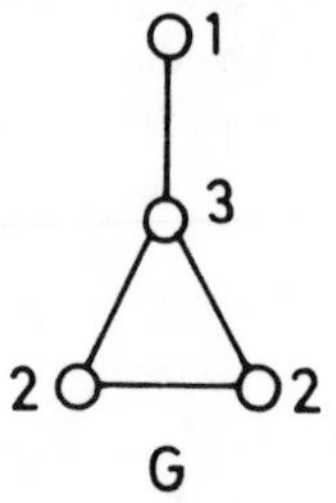

$$\sum_{i=1}^{4} D(i) = 1 + 3 + 2 + 2 = 2 \cdot 4$$

In addition, if the number of vertices of degree 1,2,3.... are denoted by F, S, T,...., the following identity is valid,

$$F + 2S + 3T + \ldots = 2M \tag{15}$$

Note,

$$F + S + T \ldots = N \tag{16}$$

Example

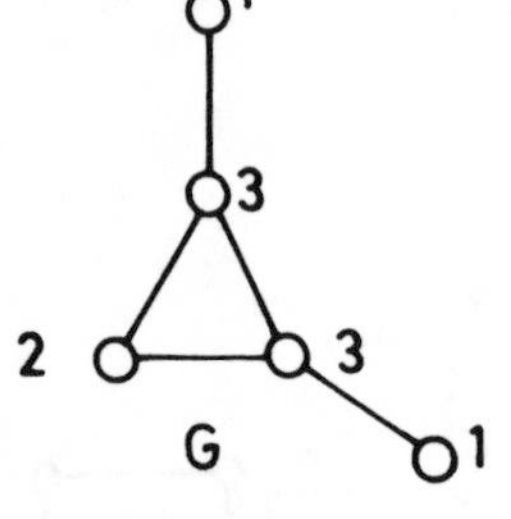

N = 5 F = 2

M = 5 S = 1

T = 2

(a) $N = F + S + T = 2 + 1 + 2 = 5$

(b) $M = F + 2S + 3T = 2 + 2 \cdot 1 + 3 \cdot 2 = 10 = 2 \cdot 5$

IV. REGULAR GRAPHS

If all vertices in a connected graph are of degree 2, the graph is called a *ring* or a *cycle*. The cycles are a special case of *regular graphs of degree D* (D = 2), which are defined as graphs having all vertices of the same degree,

$$D(v_1) = D(v_2) = \ldots = D(v_N) = D \tag{17}$$

For a regular graph of degree D,

$$M = \frac{1}{2} ND \tag{18}$$

The meaning of relation (18) is that a regular graph exists only if N and/or D are even.

If D = 1, there exists only one connected regular graph. This is a graph consisting of two vertices joined by a single edge, called *complete graph of degree one,* and denoted by K_2: ○——○. A regular graph G with N vertices and D = N − 1 is called a *complete graph* and is denoted by K_N. In such a graph all pairs of vertices are connected by adjacent edges and therefore,

$$M = \binom{N}{2} = \frac{N(N-1)}{2} \tag{19}$$

Example

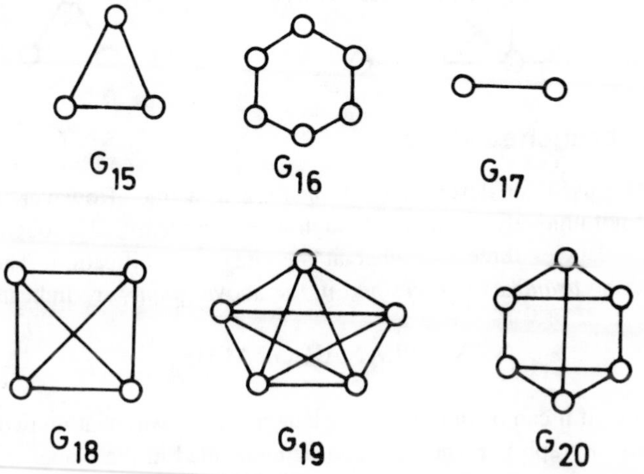

Graphs G_{15}, G_{16}, G_{17}, G_{18}, G_{19}, and G_{20} are all regular graphs. G_{15} and G_{16} are cycles with 3 and 6 vertices, respectively. G_{15}, G_{17}, G_{18}, and G_{19} are complete graphs K_3, K_2, K_4, and K_5, respectively.

V. TREES

Trees are very simple graphs. A *tree* is a connected acyclic graph. A graph is *acyclic,* if it has no rings. If a tree has N vertices, it contains N-1 edges. Obviously, by deleting any of its edges, a tree can be partitioned into two parts. A *rooted tree* is a tree in which one vertex has been distinguished from others. This vertex is usually called the *root-vertex* or simply the *root.*

Example

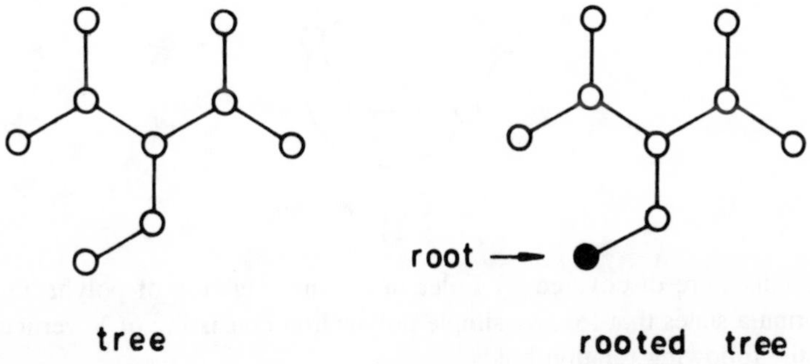

A tree necessarily possesses vertices of degree one. These are end vertices and are called *terminals*. The tree with the minimal number of terminals (two of them) is the *chain*, while the tree with the maximal number, N-1, of terminals is the *star*.

Example

chain

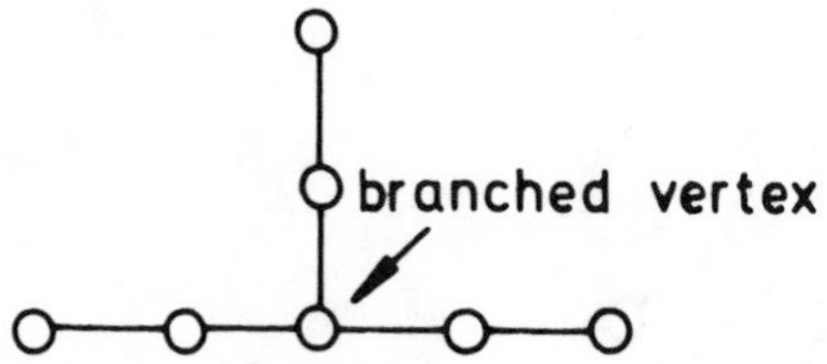

branched tree

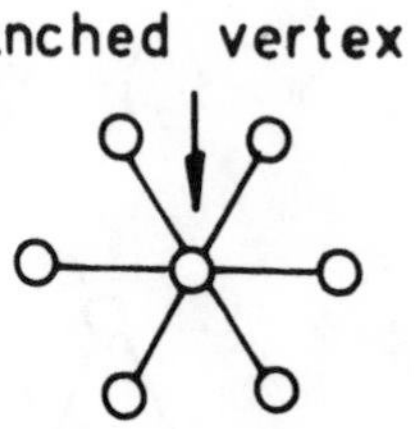

star

The above may be used for describing a *branching* in a tree. However, branching is an intuitive concept not uniquely defined,[15] though it can be identified through the appearance of vertices with valencies three or higher at the sites of ramification. These vertices are sometimes called the *branched vertices* and in the above graphs are indicated by arrows.

VI. PLANAR GRAPHS

A graph is *planar* if it can be drawn in the plane in such a way that no two edges intersect or, in other words, a graph is planar if it can be embedded in the plane.

Example

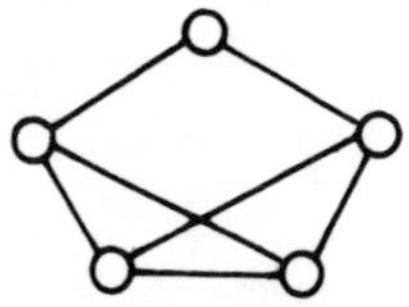

G_{21}

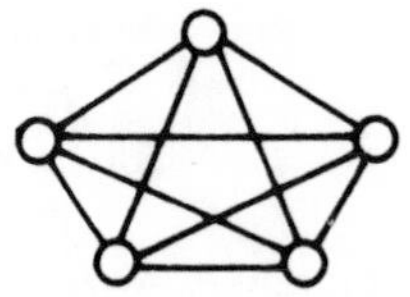

G_{19}

Graph G_{19} is nonplanar, while G_{21} is a planar graph, because it can be given, in the different representation, in the plane without any two edges crossing each other,

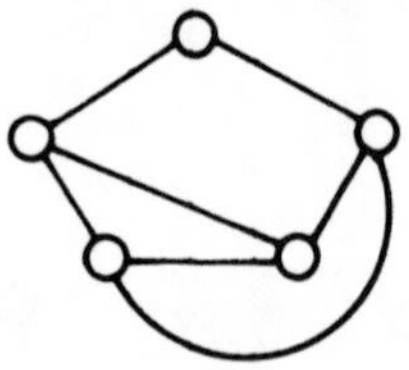

G_{21}

The planar graphs were discovered by Euler in his investigation of polyhedra. The Euler polyhedral formula states that for any simple polyhedron consisting of V vertices, E edges, and F faces, the following relation holds,

$$V - E + F = 2 \quad (20)$$

Polyhedra can be represented by *polyhedral* graphs.

Example

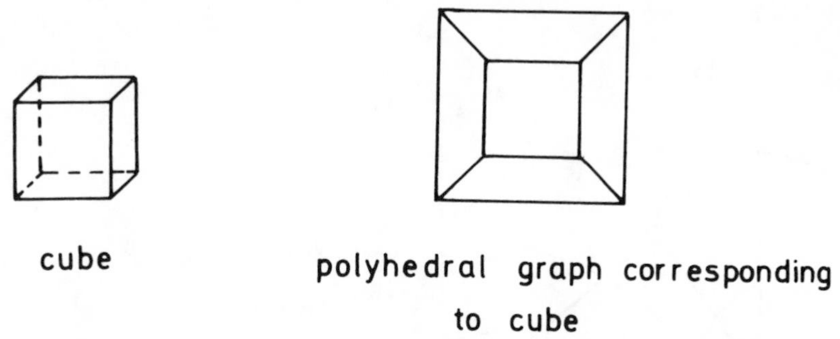

The polyhedral graphs are planar graphs for which the Euler formula holds also,

$$V(G) - E(G) + F(G) = 2 \quad (21)$$

A planar graph partitions the plane into one infinite region and several finite regions. The finite (internal) regions are called faces. The unbounded (external) regions is called infinite face. The faces count in (21) includes also the infinite face. The Euler formula can be also extended for disconnected graphs. If G is a planar disconnected graph with V(G) vertices, E(G) edges, F(G) faces, and *k* components, then the following relation is valid,

$$V(G) - E(G) + F(G) - k = 1 \quad (22)$$

Note, that in (22) the joint faces count includes only one infinite face.

Example

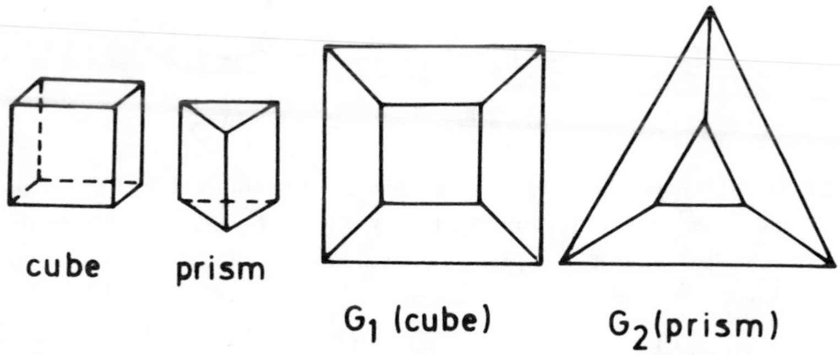

$$G = G_1 + G_2$$
$$V(G) = 8+6 = 14$$
$$E(G) = 12+9 = 21$$
$$F(G) = 5+4+1 = 10$$

$$k = V(G) - E(G) + F(G) - 1 = 14 - 21 + 10 - 1 = 2$$

Levin[16] has noticed a similarity between the formula (21) and the Gibbs phase rule,[17]

$$P + F - C = 2 \quad (23)$$

where P, F, and C denote, respectively, the number of phases, degrees of freedom, and components of the system in equilibrium between the phases. Phase diagram may be regarded as a topological simplex,[18] i.e., as the *n*-dimensional analogue of a tetrahedron, where $n > 3$. Phase diagram can be represented by a *dual graph*.[19] Given a planar graph G, its dual graph G* is constructed as follows: place a vertex in each region of G (including the exterior region) and, if two regions have an edge e in common, join the corresponding vertices by an e^* crossing only e. A dual graph of a phase diagram is constructed similarly: each field of the diagram (region of the graph) is represented as a vertex, including the field outside the diagram (the exterior region of the graph). If two fields of the phase diagram have a common edge, then the corresponding vertices are connected in the dual graph. The resulting dual graph is a multigraph.

Example

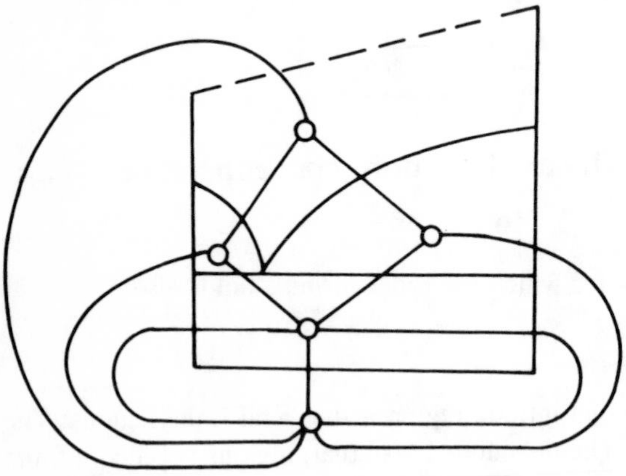

$V(G) = 5$

$E(G) = 10$

$F(G) = 7$

$V(G) - E(G) + F(G) = 5 - 10 + 7 = 2$

Depicted in this way, such diagrams are liable to computer storage and retrieval. The exact connection between the formulae (21) and (23) has never been established.

If we add new vertices on the edges of a graph G, a *homeomorph* of G is obtained.

Example

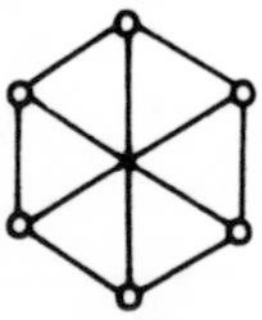

G_{22}

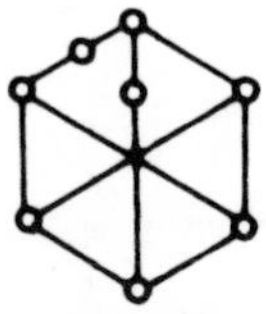

G_{23}

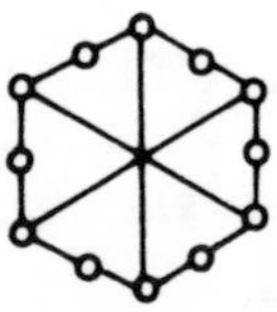

G_{24}

Graphs G_{23} and G_{24} are *homeomorphic graphs*. Two graphs are homeomorphic if they can both be obtained from the same graph by inserting new vertices of valency two into its edges. Kuratowski[20] proved that a graph is planar, if and only if, it has no subgraph homeomorphic to $K_5(= G_{19})$ or $K_{3,3}(= G_{22})$.

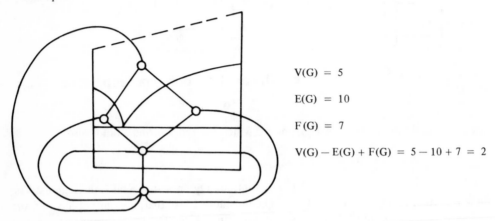

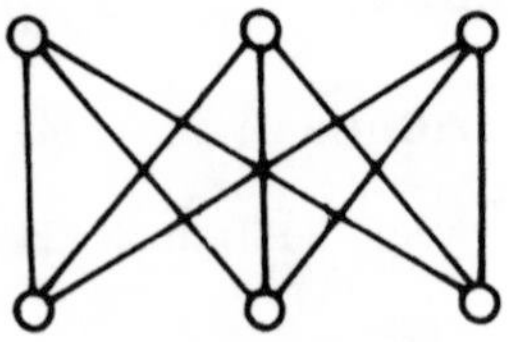

$K_{3,3}$

(For an explanation of the $K_{3,3}$ notation see Section XI of this chapter.)

VII. EULERIAN GRAPHS

At this point, we can give another definition of Eulerian graphs introduced earlier in this

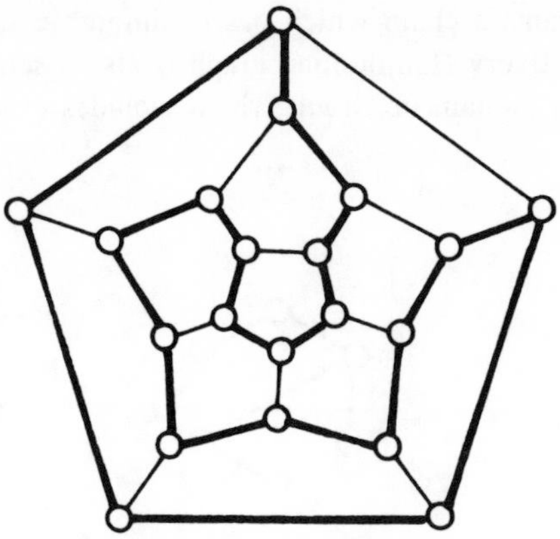

FIGURE 1. Hamilton's Hamiltonian Circuit.

chapter. A connected graph G is an Eulerian graph if, and only if, every vertex of G has *even* valency. Multigraph G_6 has all vertices with *odd* valencies and thus, it is a representative of *non-Eulerian graphs*. An Eulerian graph G contains a closed walk which includes only once every edge of G. Such a walk is called an *Eulerian trail*. If we relax the restriction that the walk must be closed, G is then a *semi-Eulerian graph*. Every Eulerian graph is also semi-Eulerian. It appears that a graph G is semi-Eulerian if, and only if, it does not contain more than two vertices of *odd* valency.

Example

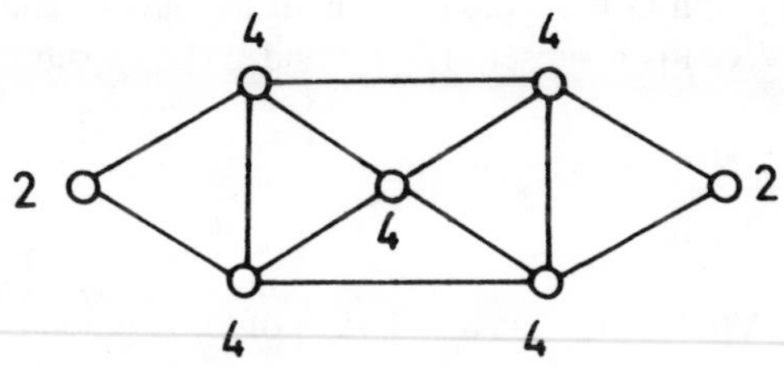

Eulerian graph

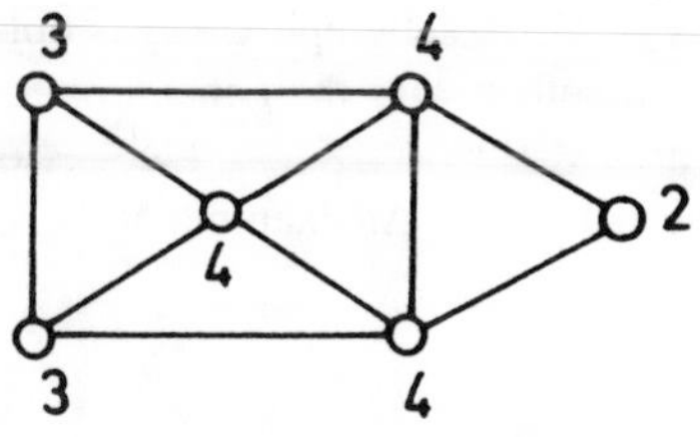

Semi-Eulerian graph

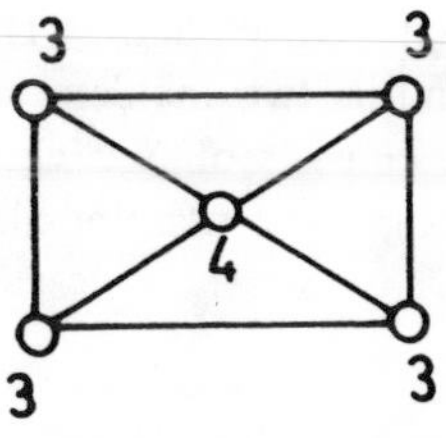

Non-Eulerian graph

VIII. HAMILTONIAN GRAPHS

A connected graph G is a *Hamiltonian graph* if it has a circuit which includes only once every vertex of G. Such a circuit is called a *Hamiltonian circuit*. A closed path is circuit. A circuit is *even* if it has an even number of edges, and *odd* otherwise.

The original Hamiltonian graph is connected with the icosian game "Around the World" invented by Hamilton (described in Biggs et al.[12]). This game uses a regular dodecahedron with 20 vertices (icosian is a Greek word signifying twenty) labeled with names of famous cities of the world, the player is challenged to "travel" around the world along the edges through each vertex exactly once. Thus, he should find a closed circuit, i.e., a Hamiltonian circuit, if he plays correctly. In Figure 1 we give Hamilton's Hamiltonian circuit.

A graph G which contains a chain which passes through every vertex of G is called a *semi-Hamiltonian graph.* Every Hamiltonian graph is also a semi-Hamiltonian graph. If a graph G does not contain a chain or circuit which includes every vertex of G, then G is called *non-Hamiltonian.*

Example

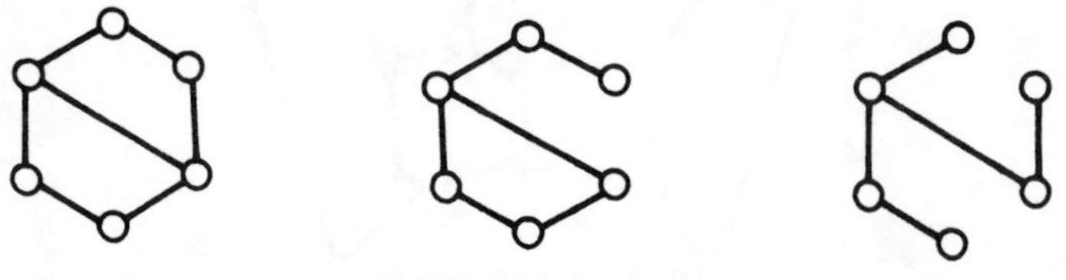

Hamiltonian graph | Semi-Hamiltonian graph | Non-Hamiltonian graph

Mallion[21] has recently demonstrated the use of semi-Hamiltonian graphs for simple ring current calculations of conjugated systems.

The determination of necessary and sufficient conditions for a graph to be Hamiltonian is still an open area in graph theory.[22]

IX. SUBGRAPHS

A subgraph G′ of a graph G is a graph which all its vertices and edges are contained in G. More formally, if V(G′) is a subset of V(G) and E (G′) a subset of E(G),

$$V(G') \subseteq V(G) \tag{24}$$

$$E(G') \subseteq E(G) \tag{25}$$

then the graph G′ = [V(G′),E(G′)] is a subgraph of G. A graph can be its own subgraph. The removal of only one edge *e* from G results in a subgraph G-*e* which consists of all vertices and M-1 edges of G. The removal of a vertex *v* from G results in a subgraph G-*v* consisting of N-1 vertices and M-D edges of G. A subgraph G-(*e*) is obtained by deletion of an edge *e* and its two incident vertices together with their incident edges from G. The subgraph G-(e) consists of N-2 vertices and M-(D′ + D″) + 1 edges of G. A spanning subgraph is a subgraph containing all the vertices of G. An example of the spanning subgraph is G-*e*.

Example

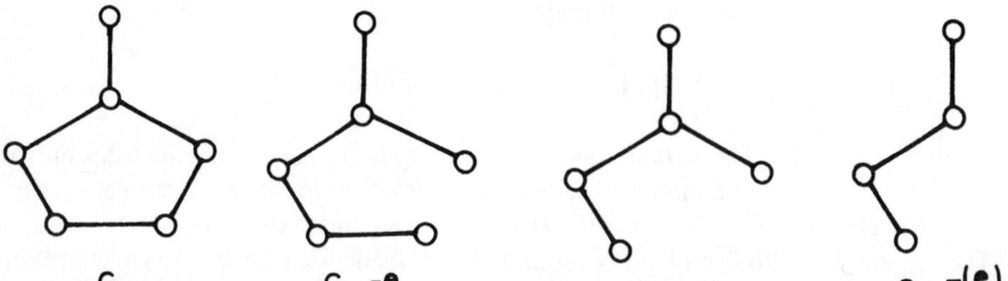

G_{25} | $G_{25}-e$ | $G_{25}-v$ | $G_{25}-(e)$

Let us form a complete set of subgraphs G-v_i, i = 1,2....,N, for a graph G′ with N ≧ 3. If the same set of subgraphs is obtained for a graph G″, the Ulam's conjecture[23] states that

G′ and G″ are isomorphic graphs. This conjecture is proved for trees[24] and for some special classes of graphs,[25] but for arbitrary graphs it remains unsolved.

X. LINE GRAPH

The *line graph* of G, denoted by L(G), is a graph derived from G in such a way that the edges in G are replaced by vertices in L(G). Two vertices in L(G) are connected whenever the corresponding edges in G are adjacent.

Example

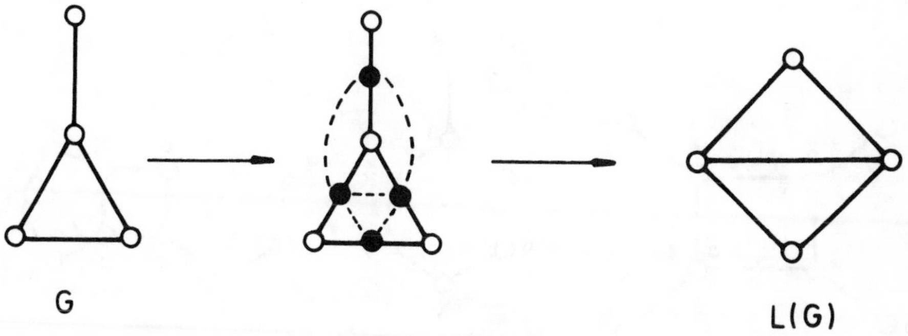

The number of vertices V(L) and the number of edges (E(L) in the line graph L(G) of G is given by the following relation,

$$V(L) = E(G) \qquad (26)$$

$$E(L) = -E(G) + ½\sum_{i} D^2(i) \qquad (27)$$

where D(i) are the degrees of the vertices in G.

Example

V(G) = 4

E(G) = 5

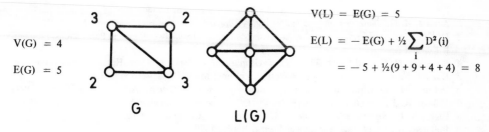

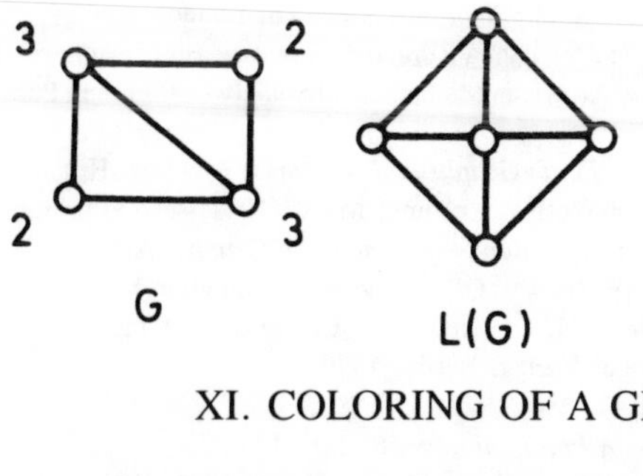

$$V(L) = E(G) = 5$$

$$E(L) = -E(G) + ½\sum_{i} D^2(i)$$

$$= -5 + ½(9 + 9 + 4 + 4) = 8$$

XI. COLORING OF A GRAPH

A coloring of a graph with a given number of colors is carried out in such a way that the adjacent vertices are always colored with different colors. If N colors are used, we have N-coloring a graph. It is clear that N-coloring of a graph with N vertices can be always carried out. If the number of colors used is gradually decreased, one finally arrives at a number *k*, called the *chromatic number,* such that *k*-coloring of a graph is possible, but not (*k*-1)-coloring. Then G is *k*-chromatic. For chemistry, the important class of graphs is *2-chromatic (bichromatic) graphs*. These graphs are also called *bipartite graphs* or *bigraphs*. The coloring of bipartite graphs will be indicated by stars (★) and circles (○). Vertices of different "colors" in bipartite graphs will be therefore called *starred* and *unstarred*. Conventionally,

$$s \geqq u \qquad (28)$$

where *s* and *u* stand for the number of starred and unstarred vertices, respectively. In bipartite graphs edges always connect vertices of different colors. This is not the case in the *non-bipartite graphs,* there will appear always an edge connecting two vertices of the same color. If a bipartite graph G is a complete bipartite graph, it is denoted as $G = K_{s,u}$. An example of a complete bigraph is the graph $K_{3,3}$.

The following theorem is given by König:[26] a graph is bipartite if, and only if, there is no odd-membered ring component of a graph. Therefore, one can decide by a simple inspection whether a graph is bipartite or nonbipartite. However, in general, the determination of *k*(G) is not possible without an effective coloring algorithm.[14]

Example

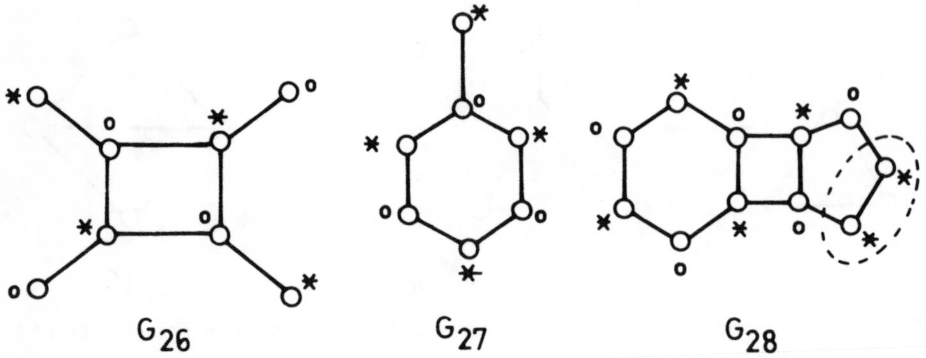

G_{26} and G_{27} are bipartite graphs, while G_{28} is a nonbipartite graph.

REFERENCES

1. **Berge, C.,** *The Theory of Graphs and Its Applications,* Methuen, London, 1962.
2. **Arnold, B. H.,** *Intuitive Concepts in Elementary Topology,* Prentice-Hall, Englewood Cliffs, N.J., 1962.
3. **Ore, O.,** *Theory of Graphs,* Vol. 38, American Mathematical Society Colloquium Publications, Providence, R.I., 1962.
4. **Bursaker, R. G. and Saaty, T. L.,** *Finite Graphs and Networks,* McGraw-Hill, New York, 1965.
5. **Harary, F.,** *Graph Theory,* Addison-Wesley, Reading, Mass., 1971, second printing.
6. **Behzad, M. and Chartrand, C.,** *Introduction to the Theory of Graphs,* Allyn & Bacon, Boston, 1971.
7. **Wilson, R. J.,** *Introduction to Graph Theory,* Oliver and Boyd, Edinburgh, 1972.
8. **Biggs, N. L.,** *Algebraic Graph Theory,* University Press, Cambridge, 1974.
9. **Bollobás, B.,** *Graph Theory,* Springer-Verlag, Berlin, 1979.
10. **Sylvester, J. J.,** *Nature (London),* 17, 284, 1877 .
11. **Euler, L.,** *Comment. Acad. Sci. Imp. Petropolitanae,* 8, 128, 1736.
12. **Biggs, N. L., Lloyd, E. K., and Wilson, R. J.,** *Graph Theory 1736—1936,* Clarendon Press, Oxford, 1976.
13. **Veblen, O.,** *Analysis Situs,* Vol. 5, 2nd ed., American Mathematical Society Colloquium Publications, New York, 1931.
14. **Zykov, A. A.,** *Teoriya konechnykh grafov,* Nauka, Novosibirsk, 1969.
15. **Essam, J. W. and Fisher, M. E.,** *Rev. Mod. Phys.,* 42, 272, 1970.
16. **Levin, I.,** *J. Chem. Educ.,* 23, 183, 1946.
17. **Moore, W. J.,** *Physical Chemistry,* 5th ed., Longman, London, 1972, 207.
18. **Kurnakov, N. S.,** *Z. Anorg. Allg. Chem.,* 169, 113, 1928.
19. **Seifer, A. L. and Stein, V. S.,** *Zh. Neorg. Khim.,* 6, 2719, 1961.
20. **Kuratowski, K.,** *Fundam. Math.,* 15, 271, 1930.
21. **Mallion, R. B.,** *Proc. R. Soc. London, Ser. A,* 341, 429, 1975.

22. **Chvátal, V.,** in *New Directions in the Theory of Graphs,* Harary, F., Ed., Academic Press, New York, 1973, 65.
23. **Ulam, S. M.,** *A Collection of Mathematical Problems,* John Wiley & Sons, New York, 1960.
24. **Kelly, P. J.,** *Pac. J. Math.,* 7, 961, 1957.
25. **Manvel, B.,** in *Proof Techniques in Graph Theory,* Harary, F., Ed., Academic Press, New York, 1969, 103.
26. **König, D.,** *Theorie der endlichen und unendlichen Graphen,* Akademische Verlagsgesellschaft, Leipzig, 1936.

Chapter 3

CHEMICAL GRAPHS

I. MOLECULAR TOPOLOGY

One property of molecules appears to be very close to a binary relation; that is two atoms in a given molecule are either bonded or not bonded. Therefore, molecules can be represented by graphs when the only property considered is the existence or not of a chemical bond. This property is called *molecular topology.*[1-3] We define molecular topology as the totality of information contained in the molecular graph. The reader should note that topology used here and later in this book is not mathematical topology. The term topology is used here, as is occasionally the case in chemical literature, in a ''weak'' sense, i.e., by recognizing vertices of valency two in molecular graphs. Thus, naphthalene and azulene in this respect are topologically nonequivalent. In a ''strong'' sense, when vertices of valency two are not discriminated, azulene and naphthalene would have the same topology, that of two fused rings (regardless of their size). The latter is sometimes referred to as the *basic topology.* The reader is reminded that topology is not a synonym for graphs. While metrics is not defined for topology, it is defined for graphs. As defined, the distance in graphs is graph theoretical invariant, but in topology there is no analogous invariant for figures and curves.

Graphs corresponding to molecules are called *chemical (molecular) graphs.* A chemical graph provides a pictorial representation of the topological structure (connectivity) of a molecule, with vertices corresponding to individual atoms and edges depicting the valence bonds between the pairs of atoms. Chemical graphs are necessarily connected graphs.

Example

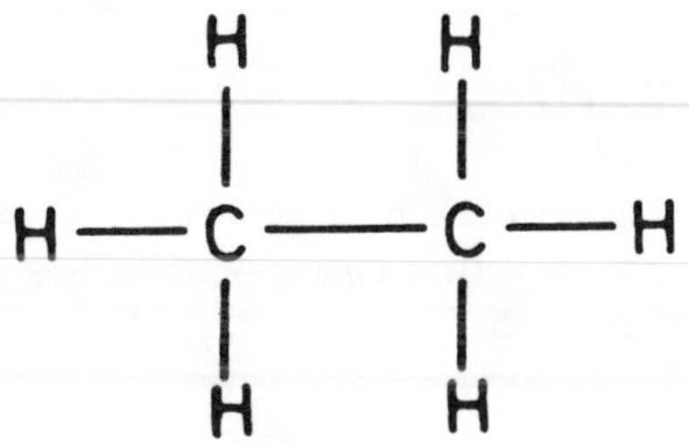

ethane

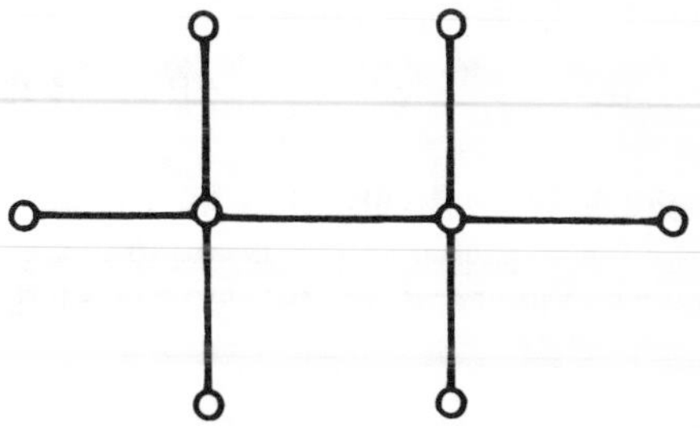

G(ethane)

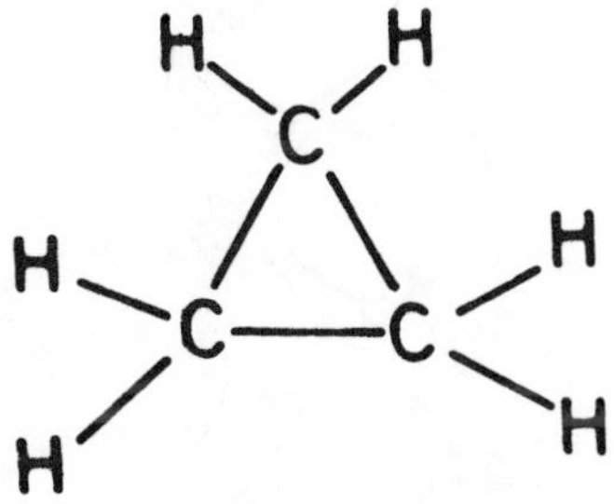

cyclopropane

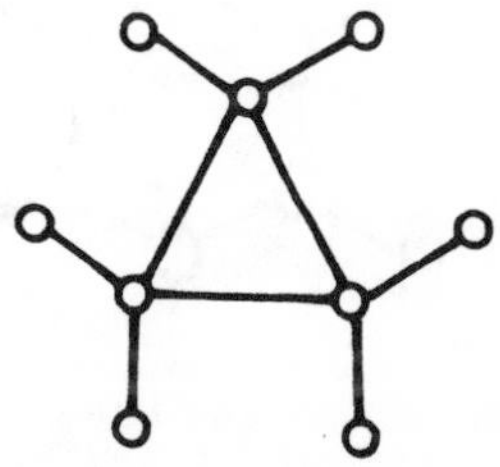

G(cyclopropane)

In order to simplify the handling of chemical graphs very often are used *hydrogen-suppressed graphs,*[4] i.e., chemical graphs depicting only molecular skeletons without considering hydrogen atoms and their bonds.

Example

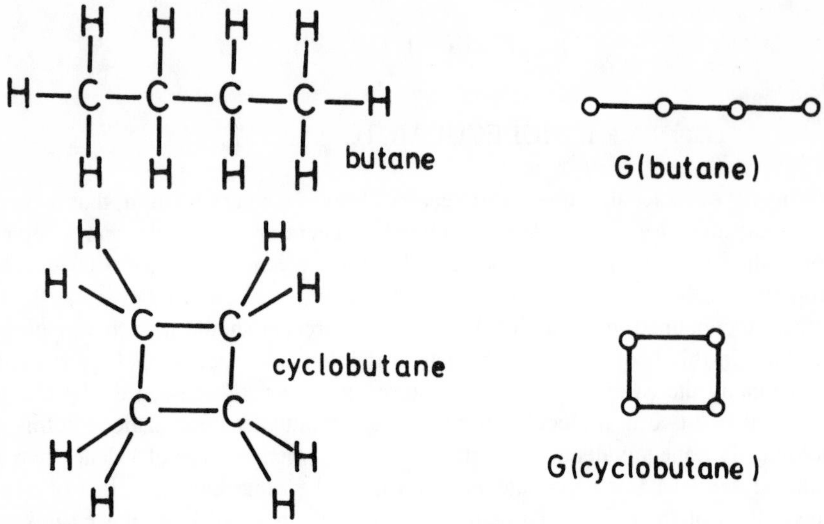

From the examples given above we see that the trees represent alkanes and the cycles cycloalkanes.

We note that the chemical graph grossly simplifies the complex picture of a molecule by depicting only its primary structure (i.e., the connections between the various pairs of atoms in the molecule) and neglecting other structural features (e.g., bond lengths, bond angles, stereochemistry, chirality). However, even so simple a picture of a molecule as a chemical graph can enable one to make useful predictions about the physical and chemical properties of molecules as we will demonstrate in this book. Since the predictions of properties and reactivity of molecules is a matter of prime interest to chemists, the development of chemical graph theory is, thus, justified.

II. HÜCKEL GRAPHS

Conjugated molecules may be conveniently represented by hydrogen-suppressed undirected planar graphs. There is one-to-one correspondence between the π-network of a given conjugated molecule and the corresponding chemical graph.

Example

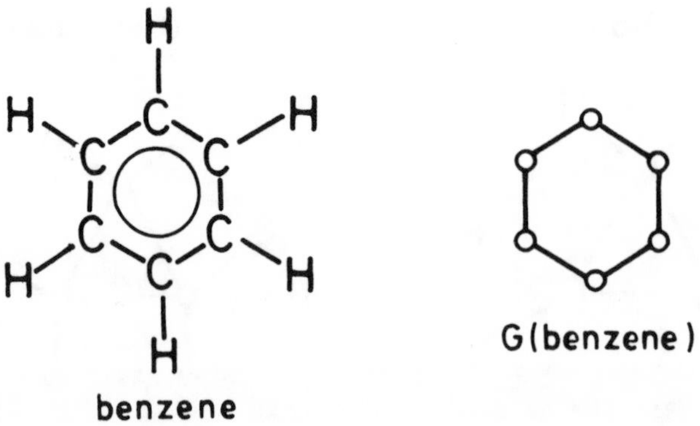

Graphs corresponding to conjugated molecules are sometimes called *Hückel graphs*.[5] Hückel graphs are undirected planar connected graphs with the maximal topological valency of vertices D_{max} = 3. Note, that we will sometimes refer to a graph theoretical valency as to a topological valency.

III. BENZENOID GRAPHS

Benzenoid graphs are a special class of Hückel graphs which are of great importance for chemistry because they depict benzenoid hydrocarbons. Benzenoid graphs are obtained by any combination of regular hexagons such that two hexagons have exactly one common edge or are disjoint. A benzenoid graph divides the plane into one infinite and a number of finite regions.[6] All vertices and edges which lie on the boundary of the infinite region form a *perimeter* of a graph. The *internal* vertices are those which do not belong to the perimeter. Their number is N_i and they all have the topological valency 3. Therefore, the internal vertices belong to three hexagons.

Example

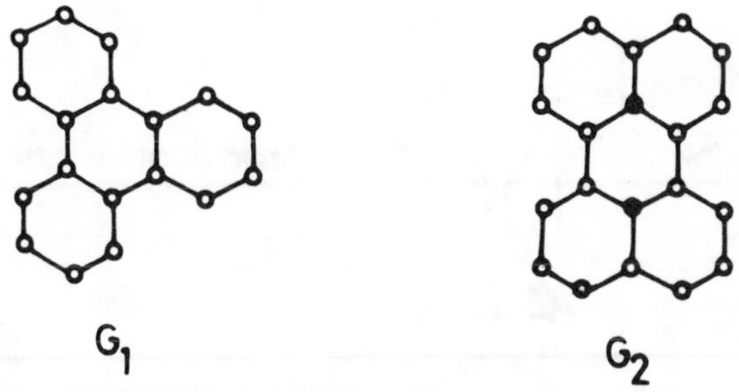

G_1 and G_2 are examples of benzenoid graphs. All their vertices lie on the perimeter, except the two internal ones in G_2 indicated by the black dots.

The following relationship between N_i, N (the total number of vertices in the benzenoid graph), and R (the number of rings in G) holds,

$$N + N_i = 4R + 2 \tag{1}$$

Example

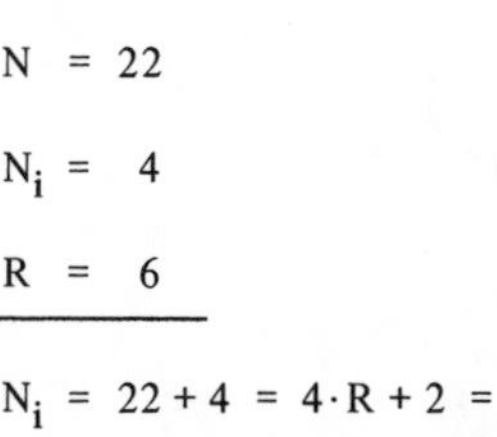

$$N = 22$$

$$N_i = 4$$

$$R = 6$$

$$N + N_i = 22 + 4 = 4 \cdot R + 2 = 26$$

If we join the centers of the adjacent pairs of hexagons in a benzenoid graph its *characteristic graph* or *inner dual* is obtained.[7] Inner dual is a subgraph of the dual graph; i.e., it does not contain the vertex corresponding to the infinite part of the plane. If inner dual graph is a tree, the benzenoid molecule is *cata-condensed,* while if it is not a tree, the benzenoid molecule is *peri-condensed.*

Example

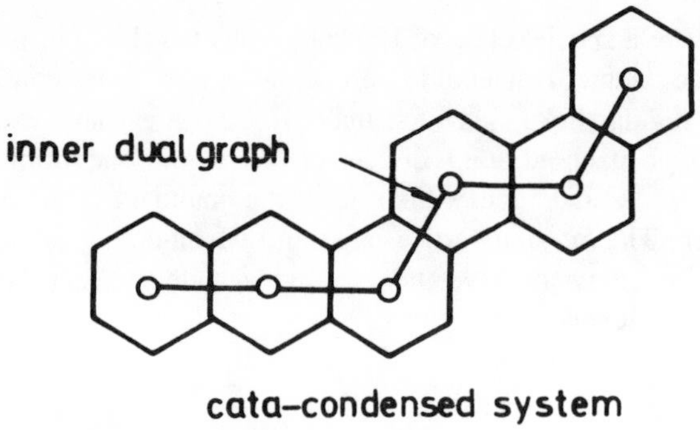

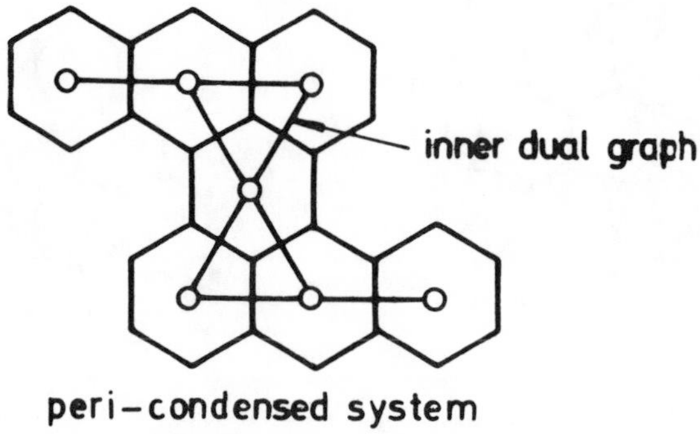

These definitions may be introduced in many alternative ways. For example, a benzenoid molecule is called cata-condensed if the number of internal vertices equals zero, or if there are no three hexagons mutually adjacent, or if all the vertices belong to the perimeter, etc.

If the inner dual graph is a cycle, the corresponding benzenoid molecule is *corona-condensed* or *cyclic cata-condensed.*[8] Circulenes or corannulenes belong to this type of benzenoids.[9-11] The number of vertices in the cycles formed from the corona-condensed benzenoid hydrocarbons must always be greater than six. Actually the smallest possible cycle which may appear as an inner dual graph of the corona-condensed structures must contain *eight* vertices.

Example

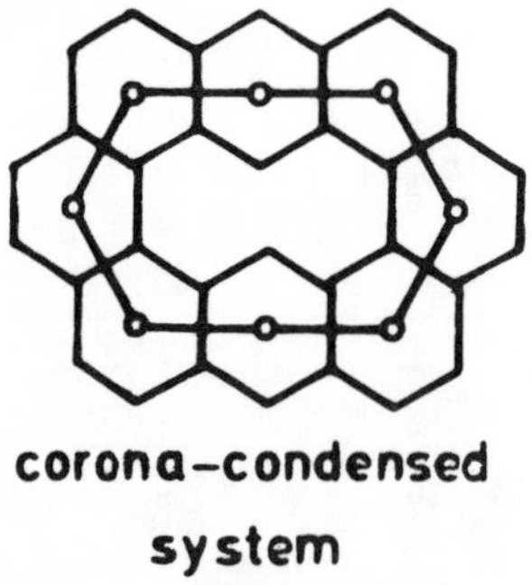

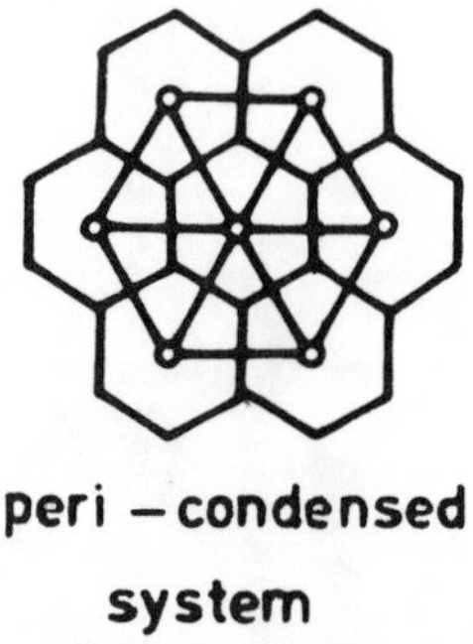

Another class of molecular graphs belonging to benzenoid systems depicts the adjacency relationship for the formal CC double bonds in individual Kekulé valence structures.[12] These graphs describe the transformation of conjugated molecules to the subspace of their double bonds. In these graphs, vertices represent double bonds and edges the coupling of double bonds.

Example

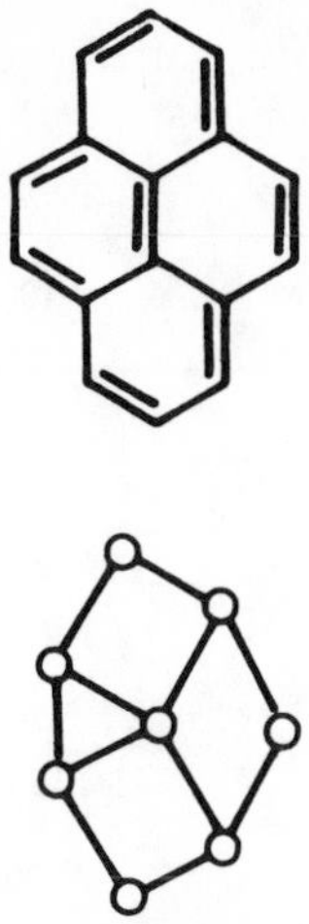

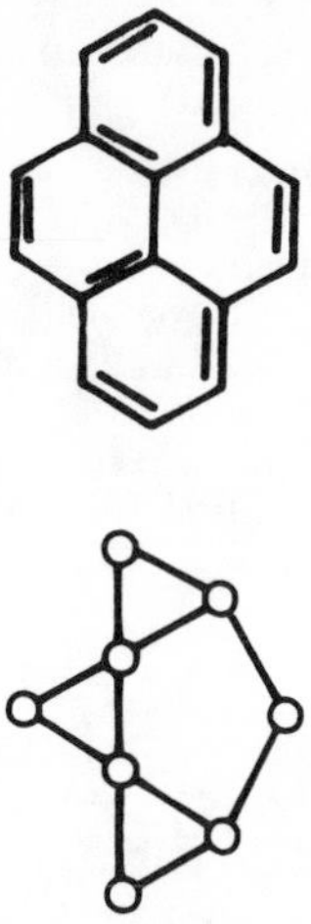

Subspace graphs dramatically illustrate the differences between individual Kekulé structures. A practical aspect of Kekulé subspace graphs is their usage in the interpretation of electronic spectra of ions of molecules with Kekulé structures.[12]

IV. KEKULÉ GRAPHS

Most conjugated systems have an alternation of single and double bonds in the structure. Conjugated molecules may have either only one possible such structure or several. It is customary to call a given structure with alternating single and double bonds the Kekulé valence structure,[13] after August Kekulé von Stradonitz who was the first to propose a cyclic structural formula for benzene,[14]

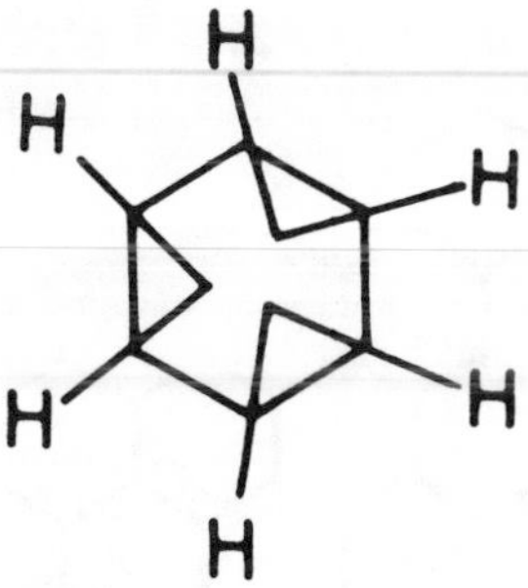

Kekulé formula of benzene (1865)

However, there are systems for which a Kekulé structural formula cannot be written. They are named *non-Kekuléan* molecules.

Example

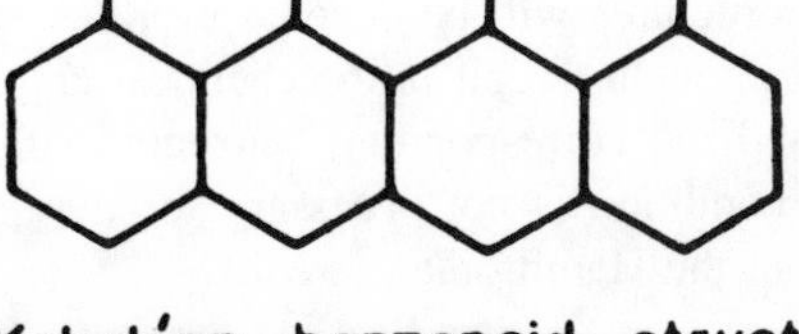

Non-Kekuléan benzenoid structure

For smaller benzenoid structures a simple criterion may be devised to predict if a molecule has Kekulé structures or not.[15] Benzenoid graphs are necessarily bipartite, that is, their vertices may be separated in a unique way into two groups: starred s and unstarred u, respectively, such that s and u are never connected. Note that $s + u = N$. If we define a number A as follows,

$$A = s - u \tag{2}$$

$$A = 0 \tag{3}$$

is a necessary condition for the existence of Kekulé structures.

Example

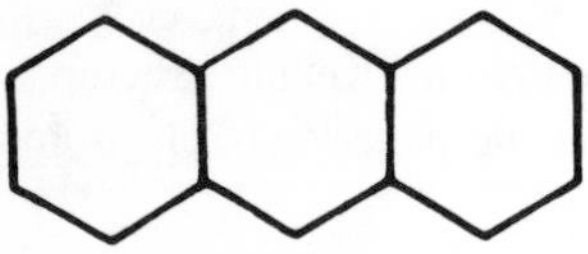

A=0

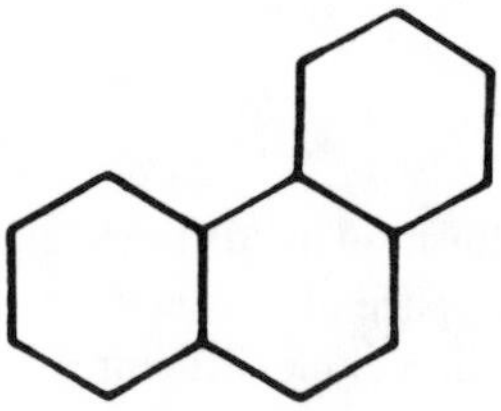

A = 0

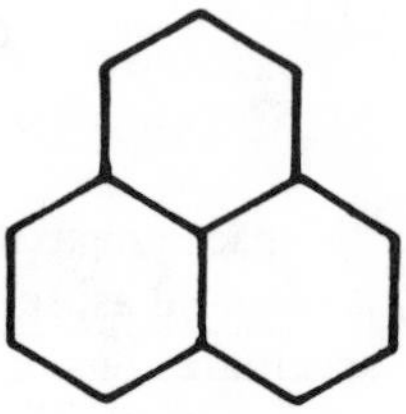

A=1

However, the condition A = 0 is not a sufficient condition for the existence of Kekulé structures, because one can find benzenoid graphs with A = 0 which do not possess Kekulé structures.

Example

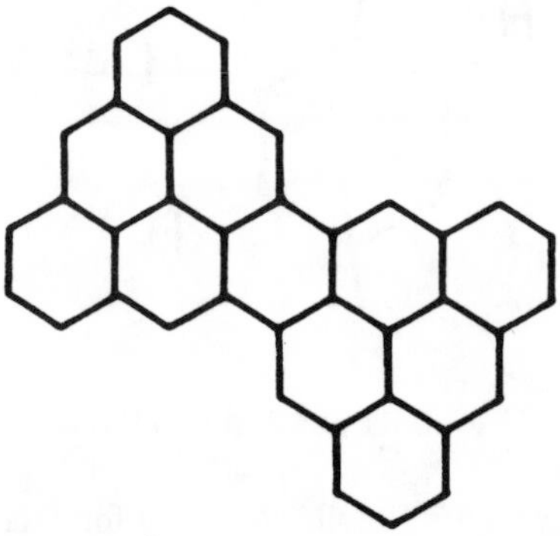

Non–Kekuléan benzenoid with A=0

It appears that non-Kekuléan benzenoids with A = 0 require at least 11 rings to show this property.

The number of Kekulé structures will be denoted by K. A sufficient condition for $K \neq 0$ is the existence of a Hamiltonian circuit in the chemical graph. If $N_i = 0$, the perimeter is a Hamiltonian circuit and the corresponding benzenoid structure must possess Kekulé structures. However, this condition is not necessary, because there are benzenoid hydrocarbons with $K \neq 0$ without the Hamiltonian circuits.

Example

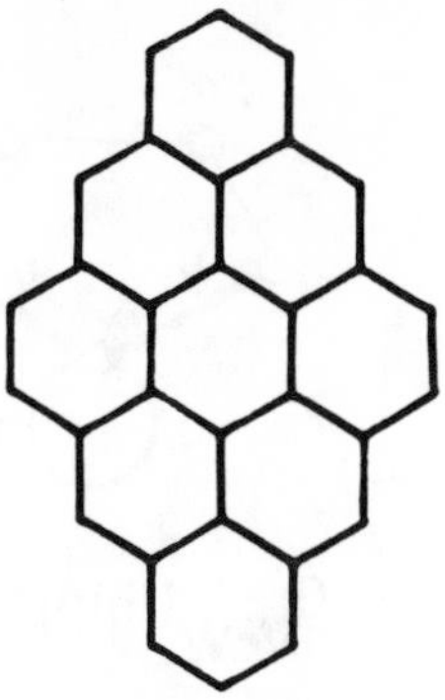

benzenoid hydrocarbon with K = 20 and without the Hamiltonian circuit

This problem, whether K = 0 or not, is brought to the reader's attention in order to emphasize that there is *no* simple recipe available to decide whether a conjugated molecule possesses Kekulé structures or not.[15a]

Kekulé structures of conjugated molecules may be represented by *Kekulé graphs*[16] or Kekuléan graphs as Sylvester called them.[17] In these graphs the binary relation corresponds to a coupling of π-orbitals forming a localized double bond. Hence, the Kekulé graphs are disconnected graphs consisting of two or more K_2 components.

Example

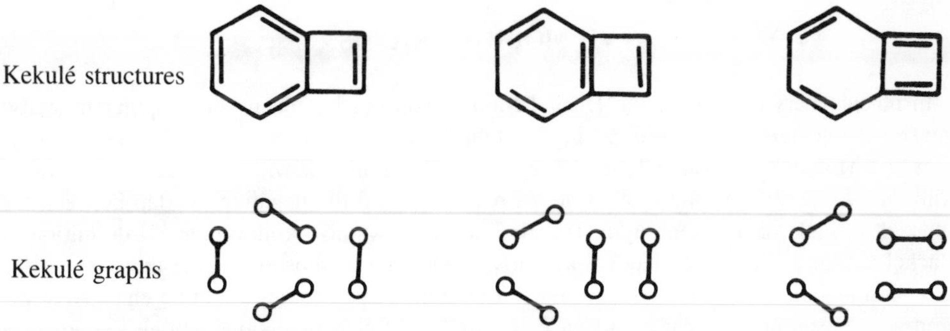

A Kekulé graph of a graph G is a spanning subgraph of G whose components are only K_2. In formal graph theory, a Kekulé graph is called a *1-factor.*[18] An n-factor is a regular graph of degree n. If G is the direct sum of n-factors, their union is called an *n-factorization* and G itself is n-factorable. Thus, Hückel graphs are 1-factorable, when they produce Kekulé graphs. A problem of interest is the enumeration of n-factors. Later on in this book (Chapter 2, Volume II) we will discuss the enumeration of 1-factors (Kekulé structures).

V. VERTEX- AND EDGE-WEIGHTED GRAPHS

Heteroconjugated molecules may be represented by *vertex-* and *edge-weighted graphs.*[19,20] Originally these graphs were named *rooted graphs,*[21,22] a name which is not quite appropriately used there, because these types of graphs belong to a class of trees.[23] A vertex- and edge-weighted graph, G_{VEW}, is a graph which has one or more of its vertices and edges distinguished in some way from the rest of vertices and edges. These vertices and edges of different "type" are weighted. Their weights are identified by parameters h (weighted vertices) and k (weighted edges) for heteroatoms and heterobonds.

Example

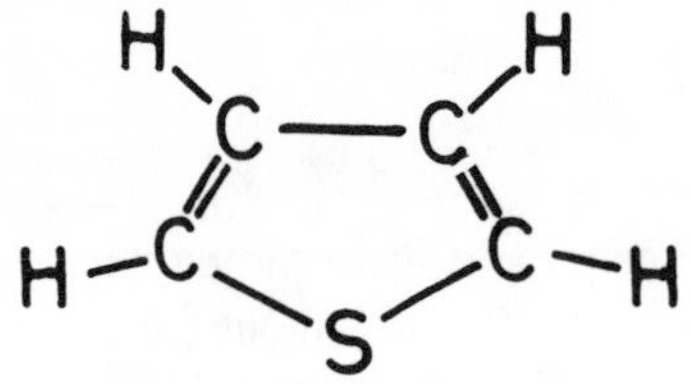

thiophen

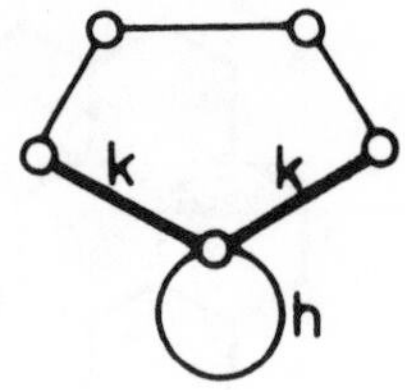

G_{VEW}(thiophen)

Weighted vertices are visually identified by loops of weight *h,* while weighted edges by heavy lines of weight *k,* respectively. Loops are sometimes called one-cycles, because they contain only one vertex.

Among the vertex- and edge-weighted graphs one may include the *quantum-chemical graphs*[24] and *reaction graphs.*[25] Quantum-chemical graphs consist of vertices representing atomic orbitals and edges representing various interactions. They belong to vertex- and edge-weighted graphs because each atomic function is represented by a differently weighted vertex and similarly, the different types of interactions are depicted by differently weighted edges. In reaction graphs, the vertices symbolize chemical species and the edges represent chemical processes involving these species.

VI. MÖBIUS GRAPHS

In recent years the use of the Hückel-Möbius concept has been very popular in studying transition states for certain pericyclic reactions, e.g., electrocyclic closures of polyenes.[26-30] Möbius systems are defined as cyclic arrays of orbitals with *one* or, more generally, with an *odd* number of phase dislocations resulting from the negative overlap between the adjacent orbitals of different sign. This definition allows also a more general definition of Hückel systems. Ordinarily Hückel systems are defined as those in which there is no sign inversion among the adjacent $2p_z$-orbitals. The sign of $2p_z$-orbitals could be changed at the arbitrary chosen atoms, leading to an *even* number of sign inversions among the adjacent orbitals. Therefore, Hückel systems may be defined as those in which there is *no* phase dislocation or in which there is an *even* number of sign inversions among the adjacent $2p_z$-orbitals.

Möbius systems may be visualized by the use of a *Möbius strip,* which was introduced by Möbius.[31] Such a strip is obtained if one short edge of a narrow sheet of paper is half-rotated and then joined with the other short edge in a band.

Möbius strip

When the $2p_z$-orbitals are drawn on the narrow sheet of paper, which is twisted into a Möbius strip with one half-twist, the Möbius system is obtained in a pictorial way. Möbius systems are also called *anti-Hückel systems,* because their stability is governed by rules opposite to those valid for Hückel systems.[32]

Möbius systems are represented by *Möbius graphs*[5,33] and are denoted by $G_{Mö}$. In mathematical literature, Möbius graphs are known as the *signed graphs*.[34] Möbius graphs should be viewed as edge-weighted graphs, $G_{Mö} = G_{EW}$. The weight of edges in these graphs is either $+1$ or -1, depending whether two adjacent $2p_z$-orbitals in a molecule are in the positive-positive or in the positive-negative overlap relationships, respectively.

Example

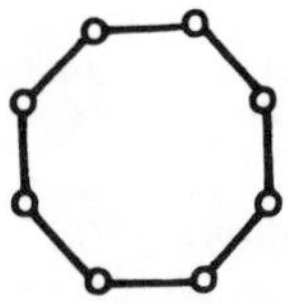

Hückel graph

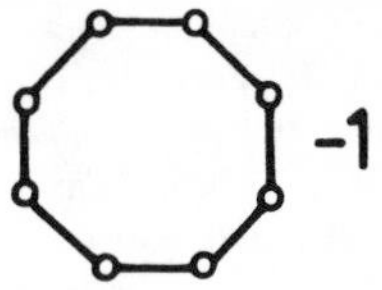

Möbius graph

The above graphs depict Hückel cyclooctatetraene and Möbius cyclooctatetraene, respectively. The placing of the connectivity -1 between the two vertices in the Möbius cyclic graph is arbitrary, but the important information is given that one phase dislocation exists in the structure called Möbius cyclooctatetraene.

In the case of Möbius graphs, there are two distinct binary relations, R^+ and R^-, but of the opposite meaning, between the connected pairs of vertices in a Möbius system. Therefore, the Möbius graph $G_{Mö}$ is in general defined as,

$$G_{Mö} = \left[V(G_{Mö}), R^+(G_{Mö}), R^-(G_{Mö})\right] \quad (4)$$

The meaning of (4) is that the two vertices in the Möbius graph either belong to R^+, if there is a $(+1)$-type of connectivity between them, or to R^-, if there is (-1)-type of connectivity between them, or they do not belong to either of these binary relations, if there is no connectivity of any kind between the vertices. A Möbius graph is fully defined when the binary relations between all pairs of vertices are defined.

REFERENCES

1. **Lederberg, J.,** *Proc. Natl. Acad. Sci. U.S.A.*, 53, 134, 1965.
2. **Gutman, I. and Trinajstić, N.,** *Topics Curr. Chem.*, 42, 49, 1973.
3. **Rouvray, D. H.,** *J. Chem. Educ.*, 52, 768, 1975.
4. **Spialter, L.,** *J. Chem. Doc.*, 4, 261, 1964; 4, 269, 1964.
5. **Graovac, A. and Trinajstić, N.,** *Croat. Chem. Acta*, 47, 95, 1975.
6. **Cvetković, D., Gutman, I., and Trinajstić, N.,** *J. Chem. Phys.*, 61, 2700, 1974.
7. **Balaban, A. T. and Harary, F.,** *Tetrahedron*, 24, 2505, 1968.
8. **Polansky, O. E. and Rouvray, D. H.,** *Math. Chem. (Mülheim/Ruhr)*, 2, 63, 1976.
9. **Hellwinkel, D.,** *Chem.-Ztg.*, 94, 715, 1970.
10. **Dopper, J. H. and Wynberg, H.,** *J. Org. Chem.*, 40, 1957, 1975.
11. **Agranat, I., Hess, B. A., Jr., and Schaad, L. J.,** *Pure Appl. Chem.*, 52, 1399, 1980.
12. **Joela, H.,** *Theor. Chim. Acta*, 39, 241, 1975.
13. **Noller, C. R.,** *Chemistry of Organic Compounds*, W. B. Saunders, Philadelphia, 1952, 406, second printing.
14. **Kekulé, A.,** *Bull. Soc. Chim. Fr.*, 3, 98, 1865; *Justus Liebigs Ann. Chem.*, 137, 129, 1866.
15. **Gutman, I.,** *Croat. Chem. Acta*, 46, 209, 1974.

15a. **Džonova-Jerman-Blažič, B. and Trinajstić, N.,** Computers & Chemistry, 6, 121, 1982.

16. **Cvetković, D., Gutman, I., and Trinajstić, N.,** *Chem. Phys. Lett.*, 16, 614, 1972.
17. **Sylvester, J. J.,** *Am. J. Math.*, 1, 64, 1878.

18. **Harary, F.,** *Graph Theory,* Addison-Wesley, Reading, Mass., 1971, 84; second printing.
19. **Graovac, A., Polansky, O. E., Trinajstić, N., and Tyutyulkov, N.,** *Z. Naturforsch.,* 30a, 1696, 1975.
20. **Rigby, M. J., Mallion, R. B., and Day, A. C.,** *Chem. Phys. Lett.,* 51, 178, 1977; erratum *Chem. Phys. Lett.,* 53, 418, 1978.
21. **Mallion, R. B., Schwenk, A. J., and Trinajstić, N.,** *Croat. Chem. Acta,* 46, 71, 1974; **Mallion, R. B., Trinajstić, N., and Schwenk, A. J.,** *Z. Naturforsch.,* 29a, 1481, 1974.
22. **Mallion, R. B., Schwenk, A. J., and Trinajstić, N.,** in *Recent Advances in Graph Theory,* Fiedler, M., Ed., Academia, Prague, 1975, 345.
23. **Harary, F. and Palmer, E. M.,** *Graphical Enumeration,* Academic Press, New York, 1973, 51.
24. **Polansky, O. E.,** *Math. Chem. (Mülheim/Ruhr),* 1, 183, 1975.
25. **Balaban, A. T., Fărcaşiu, D., and Bănică, R.,** *Rev. Roum. Chim.,* 11, 1025, 1966.
26. **Zimmerman, H. E.,** *Acc. Chem. Res.,* 4, 272, 1971.
27. **Shen, K.-W.,** *J. Chem. Educ.,* 50, 238, 1973.
28. **Smith, W. B.,** *Molecular Orbital Methods in Organic Chemistry: HMO and PMO,* Marcel Dekker, New York, 1974.
29. **Zimmerman, H. E.,** *Quantum Mechanics for Organic Chemists,* Academic Press, New York, 1975.
30. **Yates, K.,** *Hückel Molecular Orbital Theory,* Academic Press, New York, 1978, 294.
31. **Möbius, A. F.,** *Ber. K. Säches. Wiss. Leipzig Math.-Phys. Cl.,* 17, 31, 1865.
32. **Smith, W. B.,** *Molecular Orbital Methods in Organic Chemistry: HMO and PMO,* Marcel Dekker, New York, 1974, 135.
33. **Graovac, A. and Trinajstić, N.,** *J. Mol. Struct.,* 30, 416, 1976.
34. **Roberts, F. S.,** in *Applications of Graph Theory,* Wilson, R. J. and Beineke, L. W., Eds., Academic Press, London, 1979, 255.

Chapter 4

TOPOLOGICAL MATRICES

Graphs, adequately labeled, may be associated with several topological matrices.[1] A graph G is *labeled* if certain numbering of vertices of G is introduced. Here we will discuss the adjacency matrix, the incidence matrix, the circuit (cycle) matrix, and the distance matrix, respectively. A proper handling of these matrices may be used for identifying certain properties of graphs, which would not emerge ordinarily by only graph manipulations. Among the topological matrices the adjacency matrix is of the greatest importance for chemistry.[2,3]

I. THE ADJACENCY MATRIX

The most important matrix representation of a graph G is the (vertex) *adjacency matrix* **A** = **A**(G). The adjacency matrix **A**(G) of a labeled graph G with N vertices is the square N × N symmetric matrix which contain information about the internal connectivity of vertices in G. It is defined as,

$$A_{ij} = \begin{cases} 1 & \text{if, and only if, } (i,j) \in E(G) \\ 0 & \text{otherwise} \end{cases} \tag{1}$$

$$A_{ii} = 0 \tag{2}$$

A nonzero entry appears in **A**(G) only if an edge connects vertices *i* and *j*.

Example

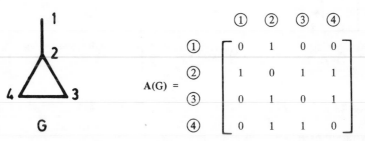

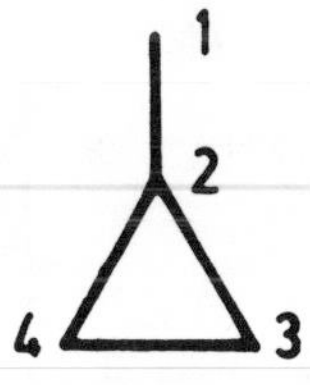

G

$$A(G) = \begin{array}{c} \\ ① \\ ② \\ ③ \\ ④ \end{array} \begin{array}{c} \begin{array}{cccc} ① & ② & ③ & ④ \end{array} \\ \begin{bmatrix} 0 & 1 & 0 & 0 \\ 1 & 0 & 1 & 1 \\ 0 & 1 & 0 & 1 \\ 0 & 1 & 1 & 0 \end{bmatrix} \end{array}$$

The adjacency matrix is symmetrical about the principal diagonal. Therefore, the transpose of the adjacency matrix $\mathbf{A}^T(G)$ leaves the adjacency matrix unchanged.

$$A^T(G) = A(G) \tag{3}$$

The transpose of a matrix $\mathbf{A}^T$ is formed by interchanging rows and columns of the matrix **A**.

For the vertex- and edge-weighted graphs, (1) and (2) should be modified,[4-6]

$$A_{ij} = \begin{cases} 1 & \text{if, and only if, } (i,j) \in E(G_{VEW}) \\ k & \text{if, and only if, } (i,j) \in E(G_{VEW}) \\ & \text{the edge } i\text{-}j \text{ is } k\text{-weighted} \\ 0 & \text{otherwise} \end{cases} \tag{4}$$

$$A_{ii} = \begin{cases} h & \text{if there is a loop of a weight} \\ & h \text{ at vertex } i \text{ in } G_{VEW} \\ 0 & \text{otherwise} \end{cases} \tag{5}$$

Example

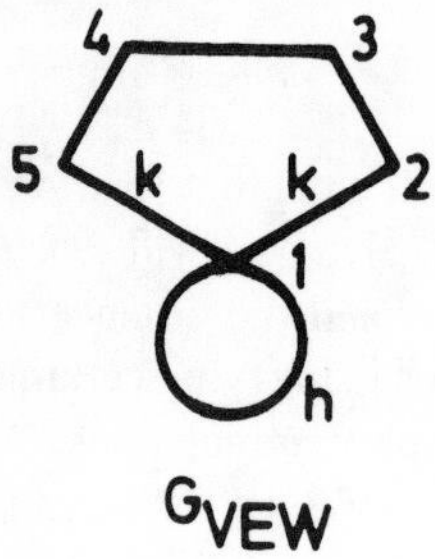

$$A(G_{VEW}) = \begin{array}{c} ① \\ ② \\ ③ \\ ④ \\ ⑤ \end{array} \overset{\begin{array}{ccccc} ① & ② & ③ & ④ & ⑤ \end{array}}{\begin{bmatrix} h & k & 0 & 0 & k \\ k & 0 & 1 & 0 & 0 \\ 0 & 1 & 0 & 1 & 0 \\ 0 & 0 & 1 & 0 & 1 \\ k & 0 & 0 & 1 & 0 \end{bmatrix}}$$

Relations (1) and (2) should be also modified for Möbius graphs,[7,8]

$$A_{ij} = \begin{cases} 1 \text{ if, and only if, } (i,j) \in E^{+}(G_{Mö}) \\ -1 \text{ if, and only if, } (i,j) \in E^{-}(G_{Mö}) \\ 0 \text{ otherwise} \end{cases} \tag{6}$$

$$A_{ii} = 0 \tag{7}$$

where $E^{+}(G_{Mö})$ is a nonempty set of edges with $(+1)$-weights, while $E^{-}(G_{Mö})$ is an odd-membered nonempty set of edges with (-1)-weights.

Example

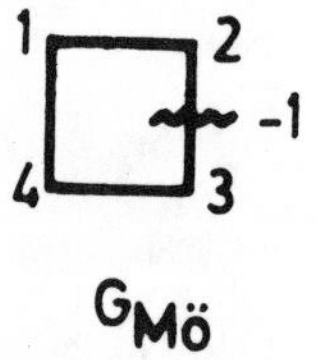

$$A(G_{Mö}) = \begin{array}{c} ① \\ ② \\ ③ \\ ④ \end{array} \overset{\begin{array}{cccc} ① & ② & ③ & ④ \end{array}}{\begin{bmatrix} 0 & 1 & 0 & 1 \\ 1 & 0 & -1 & 0 \\ 0 & -1 & 0 & 1 \\ 1 & 0 & 1 & 0 \end{bmatrix}}$$

The (edge) adjacency matrix of a graph, ${}^{E}\mathbf{A}(G)$, determines the adjacencies of edges in G. It is very rarely used. The (edge) adjacency matrix is defined as,

$${}^{E}A_{ij} = \begin{cases} 1 \text{ if, and only if, edges } i \text{ and } j \text{ are adjacent} \\ 0 \text{ otherwise} \end{cases} \tag{8}$$

$${}^{E}A_{ii} = 0 \tag{9}$$

Example

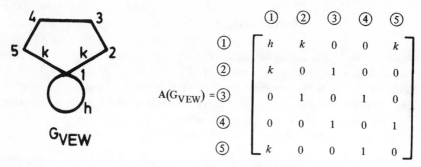

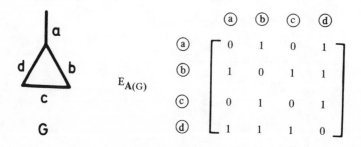

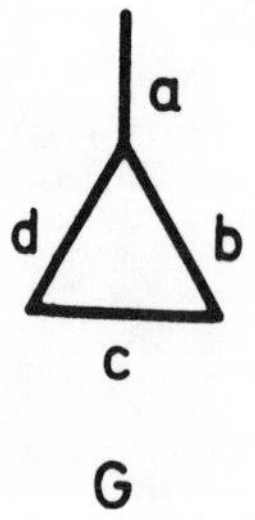

$${}^{E}\mathbf{A}(G) \quad \begin{array}{c} ⓐ \\ ⓑ \\ ⓒ \\ ⓓ \end{array} \overset{\begin{array}{cccc} ⓐ & ⓑ & ⓒ & ⓓ \end{array}}{\begin{bmatrix} 0 & 1 & 0 & 1 \\ 1 & 0 & 1 & 1 \\ 0 & 1 & 0 & 1 \\ 1 & 1 & 1 & 0 \end{bmatrix}}$$

Although the (vertex) adjacency matrix and the (edge) adjacency matrix reflect the topology of a molecule, they differ in their structure. However, the (edge) adjacency matrix of a graph G is identical to the (vertex) adjacency matrix of the line graph of G, L(G). This must be so because the edges in G are replaced by vertices in L(G) as discussed in Chapter 2, Section X.

Example

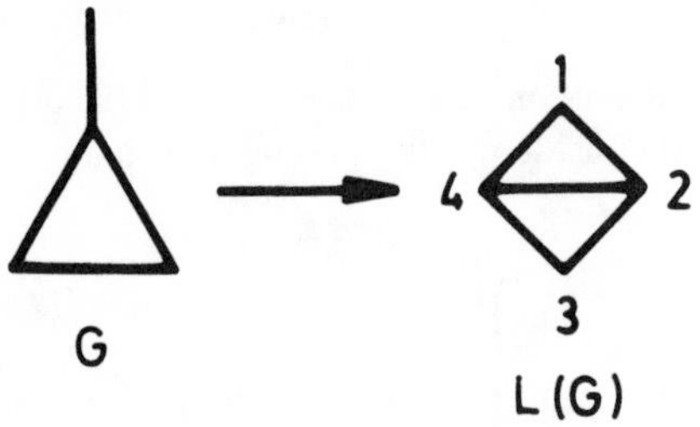

$$A[L(G)] = \begin{array}{c} \\ (1) \\ (2) \\ (3) \\ (4) \end{array} \begin{array}{c} \begin{array}{cccc} (1) & (2) & (3) & (4) \end{array} \\ \begin{bmatrix} 0 & 1 & 0 & 1 \\ 1 & 0 & 1 & 1 \\ 0 & 1 & 0 & 1 \\ 1 & 1 & 1 & 0 \end{bmatrix} \end{array}$$

This clearly illustrates that both the vertex adjacency matrix and the edge adjacency matrix are closely related topological matrices. For some graphs these matrices are identical. For example, monocyclic graphs and their corresponding line graphs have identical structures and consequently the related matrices are identical. Note, from now on when using the term adjacency matrix we will always refer to the vertex adjacency matrix.

The actual form of the adjacency matrix depends on the numbering of the vertices. For example, though the differently labeled graphs G_1 and G_2 are clearly identical,

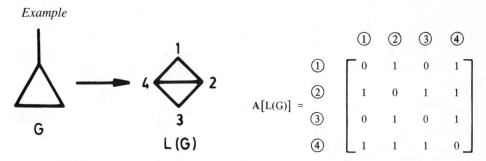

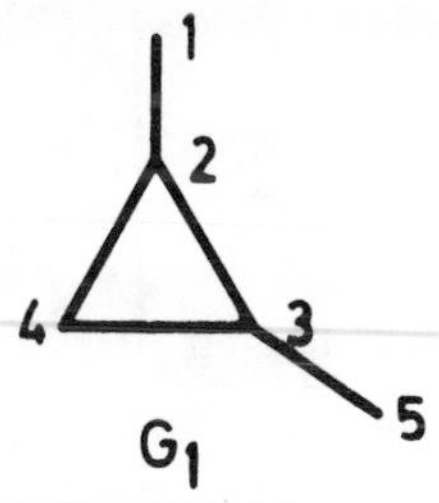

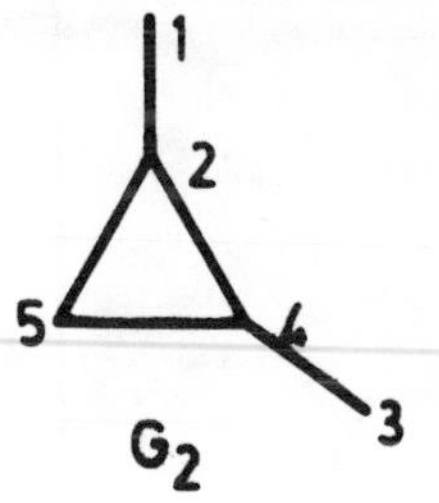

the corresponding $\mathbf{A}(G_1)$ and $\mathbf{A}(G_2)$ are not,

$$A(G_1) = \begin{bmatrix} 0 & 1 & 0 & 0 & 0 \\ 1 & 0 & 1 & 1 & 0 \\ 0 & 1 & 0 & 1 & 1 \\ 0 & 1 & 1 & 0 & 0 \\ 0 & 0 & 1 & 0 & 0 \end{bmatrix} \qquad A(G_2) = \begin{bmatrix} 0 & 1 & 0 & 0 & 0 \\ 1 & 0 & 0 & 1 & 1 \\ 0 & 0 & 0 & 1 & 0 \\ 0 & 1 & 1 & 0 & 1 \\ 0 & 1 & 0 & 1 & 0 \end{bmatrix}$$

Graphs G_1 and G_2 can be easily recognized as identical. However, a question how to recognize whether two complex graphs are identical or not, remains as one of unsolved problems of formal graph theory, although some procedures that can establish the identity of apparently different but isomorphic planar graphs have recently been suggested.[9]

The diverse numbering of vertices of a given graph can be described by means of *permutations*. The permutations P = (1,2,3,4,5) and P′ = (1,2,4,5,3) are ascribed to G_1 and G_2, respectively. If the permutation P′ is acting on the columns of $\mathbf{A}(G_1)$, i.e., the third column is placed on the position of the fifth, the fourth on the place of the third, and the fifth on the fourth, while the first and the second do not change their places, and if this is followed by the same permutation procedure on the rows, then $\mathbf{A}(G_2)$ is obtained. This procedure may be condensed in the matrix form by use of the permutation matrices. A *permutation matrix* **P** is a square N × N matrix defined as,

$$P_{ij} = \begin{cases} 1 & \text{if a mapping of vertices } i \to j \text{ is induced by permutation} \\ 0 & \text{otherwise} \end{cases} \tag{10}$$

Therefore, in each column and in each row of **P** there is one element equal to 1, and all the other elements are equal to zero. In the example considered, the permutation matrix **P** is given by,

$$\mathbf{P} = \begin{bmatrix} 1 & 0 & 0 & 0 & 0 \\ 0 & 1 & 0 & 0 & 0 \\ 0 & 0 & 0 & 1 & 0 \\ 0 & 0 & 0 & 0 & 1 \\ 0 & 0 & 1 & 0 & 0 \end{bmatrix}$$

If we multiply $\mathbf{A}(G_1)$ by **P** from the right side, the columns of $\mathbf{A}(G_1)$ will be permuted,

$$\mathbf{A}(G_1) \cdot \mathbf{P} = \begin{bmatrix} 0 & 1 & 0 & 0 & 0 \\ 1 & 0 & 0 & 1 & 1 \\ 0 & 1 & 1 & 0 & 1 \\ 0 & 1 & 0 & 1 & 0 \\ 0 & 0 & 0 & 1 & 0 \end{bmatrix}$$

The multiplication of the matrix $\mathbf{A}(G_1)\cdot\mathbf{P}$ by $\mathbf{P}^{\dagger}$ from the left side leads to the permutation of the rows with the final result summarized as,

$$\mathbf{P}^{\dagger}\,\mathbf{A}(G_1)\,\mathbf{P} = \mathbf{A}(G_2) \tag{11}$$

where $\mathbf{P}^{\dagger}$ is the adjoint of **P**. Since the permutation matrix is a unitary matrix,

$$\mathbf{P}^{\dagger}\,\mathbf{P} = \mathbf{I} \tag{12}$$

relation (11) may be rewritten as,

$$\mathbf{P}^{-1}\,\mathbf{A}(G_1)\,\mathbf{P} = \mathbf{A}(G_2) \tag{13}$$

Symbol **I** in (12) stands for a *unit matrix*.

The result stated as (13) holds generally: matrices $\mathbf{A}(G')$ and $\mathbf{A}(G'')$, corresponding to two labelings of the same graph G, are related by a similarity transformation.

Transformation (13) is called an isomorphism between graph G_1 (given by $V(G_1)$ = (1,2,3,4,5) and $E(G_1)$ = [(1,2), (2,3), (3,5), (3,4), (4,2)]) and graph G_2 (given by $V(G_2)$ = $V(G_1)$ and $E(G_2)$ = [(1,2), (2,4), (4,3), (4,5), (5,2)]). Note that G_1 and G_2 are only isomorphic graphs, but not the identical ones because $E(G_1) \neq E(G_2)$. If there exists such a transformation **P** between one and the same graph (i.e., an isomorphism of a graph with itself),

$$\mathbf{P}^{-1}\,\mathbf{A}(G)\,\mathbf{P} = \mathbf{A}(G) \tag{13a}$$

we call this transformation an *automorphism* of the graph G (actually at least one exists always: the identical one). Graph LG has only one nonidentical automorphism, namely p_1: $4\leftrightarrow4$, $2\leftrightarrow3$, $1\leftrightarrow5$. Similarly, graph G_2 has also only one nonidentical automorphism, namely p_2: $5\leftrightarrow5$, $2\leftrightarrow4$, $1\leftrightarrow3$. Clearly p_i^2 = identity. The set of all automorphisms of a graph forms a group (a group of automorphisms) which is in fact a subgroup of its vertex permutation group. Every symmetry operation of a graph corresponds to an automorphism (i.e., if there is only identical automorphism, graph has no symmetry) and vice versa, every automorphism describes some symmetry of a graph (there is one-to-one correspondence).

A. Adjacency Matrix of a Bipartite Graph

If one numbers a bipartite graph so that 1,2,...., s are starred and $s + 1$, $s + 2$,..., $s + u$ (= N) unstarred vertices, then,

$$A_{ij} = 0 \text{ for } 1 \leqq i,j \leqq s \text{ and } s + 1 \leqq i,j \leqq s + u \tag{14}$$

because two vertices of the same color are never connected. The consequence of the above is the block form of the adjacency matrix of a conveniently labeled bipartite graph,[10]

$$\mathbf{A}(G) = \begin{array}{c} \\ \star \\ \circ \end{array}\!\!\begin{array}{c} \begin{array}{cc} \star & \circ \end{array} \\ \begin{bmatrix} \mathbf{0} & \mathbf{B} \\ \mathbf{B}^T & \mathbf{0} \end{bmatrix} \end{array} \tag{15}$$

where **B** is $(s \times u)$ submatrix of **A**(G), while $\mathbf{B}^T$ is $(u \times s)$ transposed matrix of B.

Example

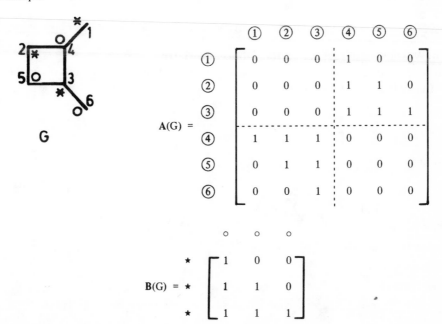

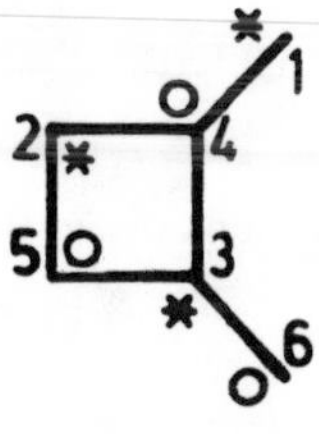

G

$$\mathbf{A}(G) = \begin{array}{c} \\ ① \\ ② \\ ③ \\ ④ \\ ⑤ \\ ⑥ \end{array}\begin{array}{c} \begin{array}{cccccc} ① & ② & ③ & ④ & ⑤ & ⑥ \end{array} \\ \left[\begin{array}{ccc|ccc} 0 & 0 & 0 & 1 & 0 & 0 \\ 0 & 0 & 0 & 1 & 1 & 0 \\ 0 & 0 & 0 & 1 & 1 & 1 \\ \hline 1 & 1 & 1 & 0 & 0 & 0 \\ 0 & 1 & 1 & 0 & 0 & 0 \\ 0 & 0 & 1 & 0 & 0 & 0 \end{array}\right] \end{array}$$

$$\mathbf{B}(G) = \begin{array}{c} \\ \star \\ \star \\ \star \end{array}\begin{array}{c} \begin{array}{ccc} \circ & \circ & \circ \end{array} \\ \begin{bmatrix} 1 & 0 & 0 \\ 1 & 1 & 0 \\ 1 & 1 & 1 \end{bmatrix} \end{array}$$

B. The Relationship Between the Adjacency Matrix and a Number of Walks in a Graph

A nonvanishing element of **A**(G), $A_{ij} = 1$, when the vertices i and j are connected, represents also a walk of a length *one* between the vertices i and j. Therefore, in general,

$$A_{ij} = \begin{cases} 1 & \text{if there is a walk of length one between vertices } i \text{ and } j \\ 0 & \text{otherwise} \end{cases} \tag{16}$$

However, there are walks of various lengths which can be found in a given graph. Thus,

$$A_{ir}\,A_{rj} = \begin{cases} 1 & \text{if there is a walk of length two between the vertices } i \text{ and } j \text{ passing through the the vertex } r \\ 0 & \text{otherwise} \end{cases} \tag{17}$$

Therefore, the expression,

$$(\mathbf{A}^2)_{ij} = \sum_{r=1}^{N} A_{ir}\,A_{rj} \tag{18}$$

represents the total number of walks of the length 2 in G between the vertices i and j. $(\mathbf{A}^2)_{ij}$ is an element of the matrix **A·A**.

For a walk of an arbitrary length ℓ, we have,

$$A_{ir}\,A_{rs}\ldots A_{zj} = \begin{cases} 1 & \text{if there is a walk of length } \ell \text{ between the vertices } i \text{ and } j \text{ passing through vertices } r,s,\ldots,z \\ 0 & \text{otherwise} \end{cases} \tag{19}$$

or

$(\mathbf{A}^{\ell})_{ij}$ = the number of walks of length ℓ between the vertices i and j

$(\mathbf{A}^{\ell})_{ij}$ is an element of the matrix $\mathbf{A}^{\ell}$. (20)

C. The Valency of a Vertex and the Adjacency Matrix

The valency of a vertex i, D(i), may be expressed as as the sum of elements in the i-th row of the adjacency matrix **A**(G),

$$D(i) = \sum_{j=1}^{N} A_{ij} \tag{21}$$

and also as,

$$D(i) = (\mathbf{A}^2)_{ii} = \sum_{j=1}^{N} A_{ij}\,A_{ji} \tag{22}$$

because,

$$A_{ij} = A_{ij}^2 = 0 \text{ or } 1 \tag{23}$$

and

$$A_{ij} = A_{ji} \tag{24}$$

There is also a connection between the trace of the adjacency matrix and the valency of a vertex. The *trace* of the square N × N matrix **S**, Tr **S**, is defined as the sum of its diagonal elements,

$$\mathrm{Tr}\,\mathbf{S} = \sum_{i=1}^{N} (\mathbf{S})_{ii} \tag{25}$$

Different labeling of the same graph results in different adjacency matrices, which are related by (13). When the cyclic property of the trace,

$$\mathrm{Tr}\,(\mathbf{ABC}) = \mathrm{Tr}\,(\mathbf{BCA}) = \mathrm{Tr}\,(\mathbf{CAB}) \tag{26}$$

is used, it follows that the trace is independent of the labeling. We write for the trace of the adjacency matrix Tr **A**. Tr **A** is a graph invariant and it is a characteristic quantity of a graph.

In the case of the adjacency matrix for graphs without loops,

$$\mathrm{Tr}\,\mathbf{A} = 0 \tag{27}$$

The trace of powers of **A** is also a graph invariant.

$$\mathrm{Tr}\,\mathbf{A}^2 = \sum_{i=1}^{N} (\mathbf{A}^2)_{ii} = \sum_{i=1}^{N} D(i) = 2M$$

$$\mathrm{Tr}\,\mathbf{A}^3 = \sum_{i=1}^{N} (\mathbf{A}^3)_{ii} = 6\,C_3 \tag{28}$$

where M is the number of edges, while C_3 is the number of three-membered cycles.

In the case of **A**(G) for graphs with *n* loops,

$$\mathrm{Tr}\,\mathbf{A} = \sum_{i=1}^{n} h_i \tag{29}$$

D. Determinant of the Adjacency Matrix

The determinant of **A**(G) will be denoted by det **A**(G). The determinant is defined by,[11,12]

$$\det \mathbf{A}(G) = \sum_{P} (-1)^P A_{1k_1} A_{2k_2} \cdots A_{Nk_N} \tag{30}$$

where the sum is over all permutations P of the integers 1,2,...,N while $(-1)^P$ is the sign of the permutation P.

It is easy to show that the value of a determinant can be found using *the method of minors:* multiply each of the elements in the first row (or column) by its minor. If a given element is in the *i*-th row and *j*-th column, the sign associated with the product is $(-1)^{i+j}$. Add all the products. The result is the value of the determinant.

Example

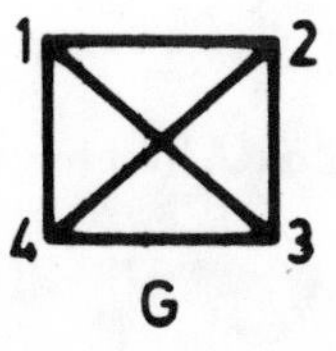

$$\mathbf{A}(G) = \begin{bmatrix} 0 & 1 & 1 & 1 \\ 1 & 0 & 1 & 1 \\ 1 & 1 & 0 & 1 \\ 1 & 1 & 1 & 0 \end{bmatrix}$$

$$\det \mathbf{A}(G) = \begin{vmatrix} 0 & 1 & 1 & 1 \\ 1 & 0 & 1 & 1 \\ 1 & 1 & 0 & 1 \\ 1 & 1 & 1 & 0 \end{vmatrix} = \begin{vmatrix} 1 & 1 & 1 \\ 1 & 0 & 1 \\ 1 & 1 & 0 \end{vmatrix} + \begin{vmatrix} 1 & 1 & 1 \\ 0 & 1 & 1 \\ 1 & 1 & 0 \end{vmatrix}$$

$$- \begin{vmatrix} 1 & 1 & 1 \\ 0 & 1 & 1 \\ 1 & 0 & 1 \end{vmatrix} = - \begin{vmatrix} 0 & 1 \\ 1 & 0 \end{vmatrix} + \begin{vmatrix} 1 & 1 \\ 1 & 0 \end{vmatrix} - \begin{vmatrix} 1 & 1 \\ 0 & 1 \end{vmatrix}$$

$$+ \begin{vmatrix} 1 & 1 \\ 1 & 0 \end{vmatrix} + \begin{vmatrix} 1 & 1 \\ 1 & 1 \end{vmatrix} - \begin{vmatrix} 1 & 1 \\ 0 & 0 \end{vmatrix} - \begin{vmatrix} 1 & 1 \\ 1 & 1 \end{vmatrix} = -3$$

The evaluation and manipulations of determinants are greatly simplified because of the following properties of determinants:[11,12]

1. The value of a determinant changes sign when two rows (or columns) are interchanged (see 3×3 determinants in the above example).
2. If every element in a row (or column) is zero, the value of the determinant is zero.
3. If two rows (or columns) are equal, the determinant is zero.
4. If all elements in a row (or column) are multiplied by a constant, the value of the determinant is multiplied by the same constant.
5. If the elements of the determinant are arranged as follows,

$$\begin{vmatrix} 1 & 2 & 3 \ldots\ldots & N \\ 2 & 3 & 4 \ldots\ldots & 1 \\ 3 & 4 & 5 \ldots\ldots & 2 \\ \ldots & \ldots & \ldots\ldots & \ldots \\ N & 1 & 2 \ldots\ldots & N-1 \end{vmatrix} \qquad (31)$$

then the value of the determinant is given by,

$$\det = \frac{1}{2}(-1)^{N(N-1)/2}\,(N+1)\,N^{N-1} \qquad (32)$$

6. The transposed adjacency matrix $\mathbf{A}^T(G)$ has the same value of the determinant as the adjacency matrix $\mathbf{A}(G)$.

E. Permanent of the Adjacency Matrix

The permanent of the adjacency matrix **A**(G) is defined as,[13]

$$\text{per } \mathbf{A}(G) = \sum_{P} A_{1k_1} A_{2k_2} \cdots A_{Nk_N} \tag{33}$$

where the summation extends over the N! permutations of the elements 1,2,….,N.

Example

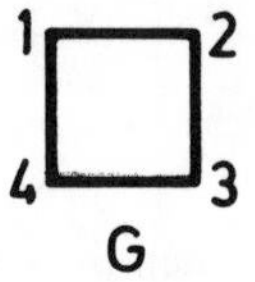

$$\mathbf{A}(G) = \begin{bmatrix} 0 & 1 & 0 & 1 \\ 1 & 0 & 1 & 0 \\ 0 & 1 & 0 & 1 \\ 1 & 0 & 1 & 0 \end{bmatrix}$$

$$\det \mathbf{A}(G) = 0$$

$$\text{per } \mathbf{A}(G) = 4$$

The permanent of the (adjacency) matrix has the following properties:[13]

1. The transposed adjacency matrix $\mathbf{A}^T(G)$ has the same value of the permanent as the adjacency matrix **A**(G),

$$\text{per } \mathbf{A}^T(G) = \text{per } \mathbf{A}(G) \tag{34}$$

2. The permanent of the adjacency matrix remains constant when any two rows and columns of **A**(G) are interchanged. The interchange of any two (or more) rows and columns of **A**(G) corresponds to relabeling of two (or more) vertices of G.

Example

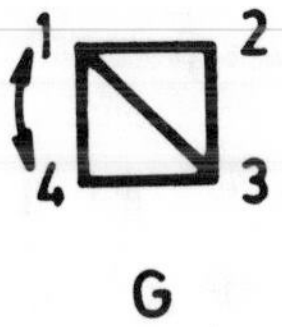

$$\mathbf{A}(G) = \begin{array}{c} \begin{array}{cccc} \textcircled{1} & \textcircled{2} & \textcircled{3} & \textcircled{4} \end{array} \\ \begin{bmatrix} 0 & 1 & 1 & 1 \\ 1 & 0 & 1 & 0 \\ 1 & 1 & 0 & 1 \\ 1 & 0 & 1 & 0 \end{bmatrix} \end{array} \begin{array}{c} \\ \textcircled{1} \\ \textcircled{2} \\ \textcircled{3} \\ \textcircled{4} \end{array}$$

$$\text{per } \mathbf{A}(G) = 4$$

$$\mathbf{A}(G') = \begin{array}{c} \begin{array}{cccc} \textcircled{1} & \textcircled{2} & \textcircled{3} & \textcircled{4} \end{array} \\ \begin{bmatrix} 0 & 0 & 1 & 1 \\ 0 & 0 & 1 & 1 \\ 1 & 1 & 0 & 1 \\ 1 & 1 & 1 & 0 \end{bmatrix} \end{array} \begin{array}{c} \\ \textcircled{1} \\ \textcircled{2} \\ \textcircled{3} \\ \textcircled{4} \end{array} \qquad \text{per } \mathbf{A}(G') = 4$$

3. The permanent of the (adjacency) matrix with a zero row and/or column is equal to zero. (The adjacency matrix with a zero row and/or column corresponds to a disconnected graph).

Example

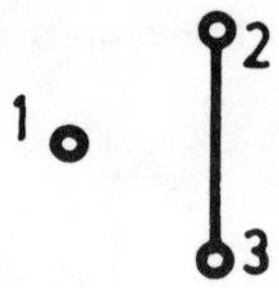

G

$$A(G) = \begin{bmatrix} 0 & 0 & 0 \\ 0 & 0 & 1 \\ 0 & 1 & 0 \end{bmatrix}$$

$$\text{per } \mathbf{A}(G) = 0$$

4. If the adjacency matrix **A**(G) has all elements equal one, then the permanent of **A**(G) is given by,

$$\text{per } \mathbf{A}(G) = N! \tag{35}$$

(The adjacency matrix of a vertex-weighted graph G_{VW} with loops of the weight one has all entries equal to 1).

Example

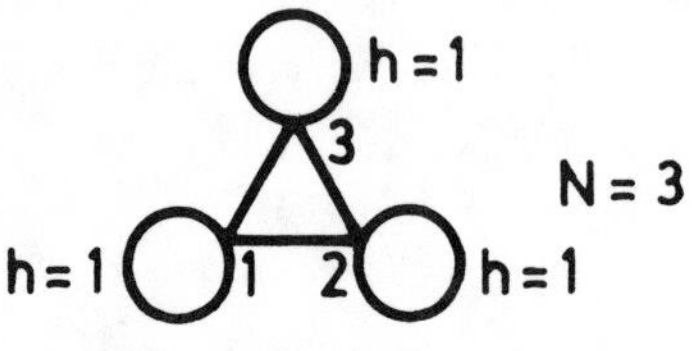

G_{VW}

$$\mathbf{A}(G_{VW}) = \begin{bmatrix} 1 & 1 & 1 \\ 1 & 1 & 1 \\ 1 & 1 & 1 \end{bmatrix}$$

$$\text{per } \mathbf{A}(G_{VW}) = 1 \cdot 2 \cdot 3 = 6$$

5. If the adjacency matrix **A**(G) has all elements equal 1, but one being zero, the permanent of **A**(G) is given as follows,

$$\text{per } \mathbf{A}^{(1)}(G) = N! - (N-1)! = (N-1)\,[(N-1)!] \tag{36}$$

(The adjacency matrix $A^{(1)}$ of a vertex weighted graph G_{VW} with $N - 1$ loops of the weight one has all entries equal 1, but one which is zero).

Example

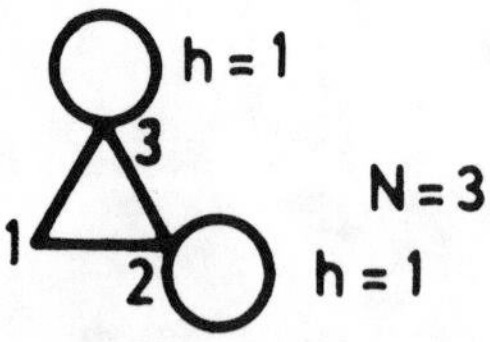

G_{VW}

$$\mathbf{A}(G_{VW}) = \begin{bmatrix} 0 & 1 & 1 \\ 1 & 1 & 1 \\ 1 & 1 & 1 \end{bmatrix}$$

$$\text{per } \mathbf{A}(G_{VW}) = 2 \cdot 1 \cdot 2 = 4$$

6. The expression (36) may be generalized for the case of the adjacency matrix with *k* zero entries $\mathbf{A}^{(k)}$, with restriction $N \geq k$,

$$\text{per } \mathbf{A}^{(k)} = N! - \binom{k}{1}(N-1)! + \binom{k}{2}(N-2)! - \ldots + (-1)^k \binom{k}{k}(N-k)! \tag{37}$$

Example

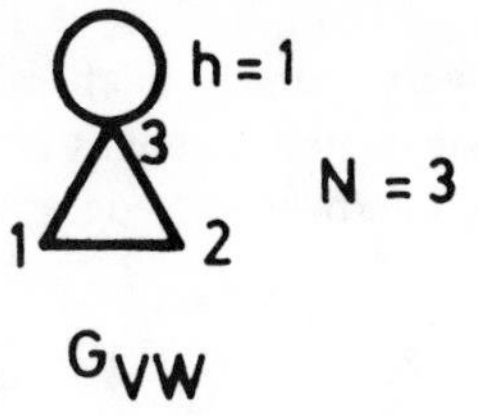

$$A^{(k)}(G_{VW}) = \begin{bmatrix} 0 & 1 & 1 \\ 1 & 0 & 1 \\ 1 & 1 & 1 \end{bmatrix} \quad ; k = 2$$

$$\text{per } \mathbf{A}^{(2)}(G_{VW}) = 3! - \binom{2}{1}(N-1)! + $$

$$+ (-1)^2 \binom{2}{2}(3-2)! = 3$$

F. The Inverse of the Adjacency Matrix

The inverse of the adjacency matrix $\mathbf{A}^{-1}(G)$ is defined by,

$$\mathbf{A}^{-1}(G)\,\mathbf{A}(G) = \mathbf{A}(G)\,\mathbf{A}^{-1}(G) = I \tag{38}$$

The inverse of the adjacency matrix may be obtained by,

$$\mathbf{A}^{-1}(G) = \frac{\text{adj } \mathbf{A}(G)}{\det \mathbf{A}(G)} \tag{39}$$

where adj $\mathbf{A}(G)$ is the matrix adjoint to $\mathbf{A}(G)$. The matrix adjoint to $\mathbf{A}(G)$ may be obtained by first replacing each matrix element $(\mathbf{A})_{ij}$ by its cofactor in the det $\mathbf{A}(G)$, and then transposing rows and columns. However, if the adjacency matrix $\mathbf{A}(G)$ has an inverse $\mathbf{A}^{-1}(G)$, the determinant of $\mathbf{A}(G)$ must not vanish,

$$\det \mathbf{A}(G) \neq 0 \tag{40}$$

Note, the minor with the sign $(-1)^{i+j}$ is called *cofactor*.

Example

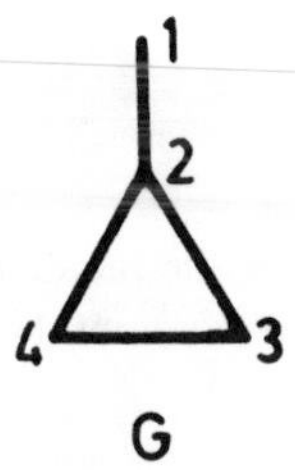

$$\mathbf{A}(G) = \begin{bmatrix} 0 & 1 & 0 & 0 \\ 1 & 0 & 1 & 1 \\ 0 & 1 & 0 & 1 \\ 0 & 1 & 1 & 0 \end{bmatrix}$$

$$\det \mathbf{A}(G) = 1$$

$$\text{adj } \mathbf{A}(G) = \begin{bmatrix} 2 & 1 & -1 & -1 \\ 1 & 0 & 0 & 0 \\ -1 & 0 & 0 & 1 \\ -1 & 0 & 1 & 0 \end{bmatrix} \quad \mathbf{A}^{-1}(G) = \begin{bmatrix} 2 & 1 & -1 & -1 \\ 1 & 0 & 0 & 0 \\ -1 & 0 & 0 & 1 \\ -1 & 0 & 1 & 0 \end{bmatrix}$$

$$\mathbf{A}(G)\cdot\mathbf{A}^{-1}(G) = \begin{bmatrix} 1 & 0 & 0 & 0 \\ 0 & 1 & 0 & 0 \\ 0 & 0 & 1 & 0 \\ 0 & 0 & 0 & 1 \end{bmatrix}$$

II. THE INCIDENCE MATRIX

The incidence matrix of a graph[1] **T**(G) has been used very sparsely in chemistry.[2,14] The incidence matrix **T**(G) of a labeled graph G with N vertices and M edges is the N × M array; the rows and columns of the matrix corresponding to vertices and edges, respectively, of G. It is defined as,

$$T_{ij} = \begin{cases} 1 & \text{if the j-th edge is incident with the i-th vertex} \\ 0 & \text{otherwise} \end{cases} \qquad (41)$$

The sum of all the matrix elements of **T**(G) is equal twice the number of edges in G,

$$\sum_{(i,j)} T_{ij} = 2M \qquad (42)$$

or consequently to the valencies of the vertices in G,

$$\sum_{(i,j)} T_{ij} = \sum_{i} D_i \qquad (43)$$

Example

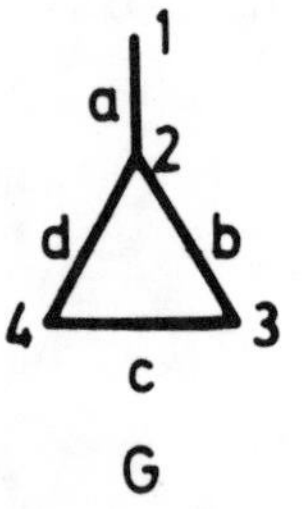

$$\mathbf{T}(G) = \begin{array}{c} \\ ① \\ ② \\ ③ \\ ④ \end{array} \begin{array}{c} \begin{array}{cccc} ⓐ & ⓑ & ⓒ & ⓓ \end{array} \\ \begin{bmatrix} 1 & 0 & 0 & 0 \\ 1 & 1 & 0 & 1 \\ 0 & 1 & 1 & 0 \\ 0 & 0 & 1 & 1 \end{bmatrix} \end{array}$$

$$\sum_{(i,j)} T_{ij} = 8 = 2 \cdot 4$$

A graph is completely determined either by the adjacency matrix or the incidence matrix. Thus, if two graphs have either of these matrices identical, they are isomorphic. The adjacency matrix and the incidence matrix are related by,[1,2,15]

$$\mathbf{A}[L(G)] = \mathbf{T}^{T}(G)\,\mathbf{T}(G) - 2\,\mathbf{I}_M \qquad (44)$$

where $\mathbf{T}^{T}(G)$ is the transpose matrix of the (N × M) incidence matrix of G, $\mathbf{I}_M$ is (M × M) unit matrix, while **A**[L(G)] stands for the adjacency matrix of the line graph of G.

Example

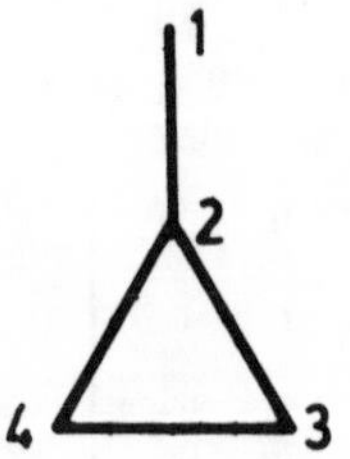

$$\mathbf{T}^T(G)\,\mathbf{T}(G) = \begin{bmatrix} 1 & 1 & 0 & 0 \\ 0 & 1 & 1 & 0 \\ 0 & 0 & 1 & 1 \\ 0 & 1 & 0 & 1 \end{bmatrix} \cdot \begin{bmatrix} 1 & 0 & 0 & 0 \\ 1 & 1 & 0 & 1 \\ 0 & 1 & 1 & 0 \\ 0 & 0 & 1 & 1 \end{bmatrix} = \begin{bmatrix} 2 & 1 & 0 & 1 \\ 1 & 2 & 1 & 1 \\ 0 & 1 & 2 & 1 \\ 1 & 1 & 1 & 2 \end{bmatrix}$$

$$-2\,\mathbf{I}_M = \begin{bmatrix} -2 & 0 & 0 & 0 \\ 0 & -2 & 0 & 0 \\ 0 & 0 & -2 & 0 \\ 0 & 0 & 0 & -2 \end{bmatrix} \qquad \mathbf{T}^T(G)\,\mathbf{T}(G) - 2\mathbf{I}_M = \begin{bmatrix} 0 & 1 & 0 & 1 \\ 1 & 0 & 1 & 1 \\ 0 & 1 & 0 & 1 \\ 1 & 1 & 1 & 0 \end{bmatrix}$$

$$\mathbf{A}[L(G)] = \begin{bmatrix} 0 & 1 & 0 & 1 \\ 1 & 0 & 1 & 1 \\ 0 & 1 & 0 & 1 \\ 1 & 1 & 1 & 0 \end{bmatrix}$$

Note, that the adjacency matrix and the incidence matrix determine a graph up to isomorphism.

III. THE CIRCUIT MATRIX

The circuit (cycle) matrix of a graph **C**(G), whose cycles (circuits) c and edges e are labeled, is a rectangular ($c \times e$) array defined as,[2]

$$C_{ij} = \begin{cases} 1 & \text{if the } i\text{-th cycle contains the } j\text{-th edge} \\ 0 & \text{otherwise} \end{cases} \tag{45}$$

Example

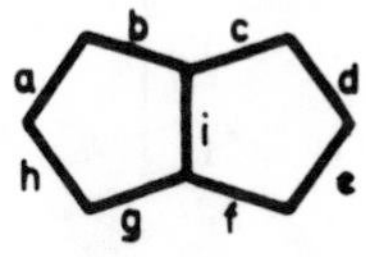

G

cycles:

$c_1 = \{a, b, i, g, h\}$

$c_2 = \{c, d, e, f, i\}$

$c_3 = \{a, b, c, d, e, f, g, h\}$

$$\mathbf{C}(G) = \begin{array}{c} \\ (c_1) \\ (c_2) \\ (c_3) \end{array} \begin{array}{c} \begin{array}{ccccccccc} (a) & (b) & (c) & (d) & (e) & (f) & (g) & (h) & (i) \end{array} \\ \begin{bmatrix} 1 & 1 & 0 & 0 & 0 & 0 & 1 & 1 & 1 \\ 0 & 0 & 1 & 1 & 1 & 1 & 0 & 0 & 1 \\ 1 & 1 & 1 & 1 & 1 & 1 & 1 & 1 & 0 \end{bmatrix} \end{array}$$

The cycle matrix of a graph G, **C**(G), is related to incidence matrix **T**(G) by the following orthogonality relationship,

$$\mathbf{C}(G)\,\mathbf{T}^T(G) \equiv \mathbf{0} (\text{mod } 2) \tag{46}$$

The orthogonality here is that of *modulo 2;* an element of the product is assumed zero if the number is divisible by 2.

Example

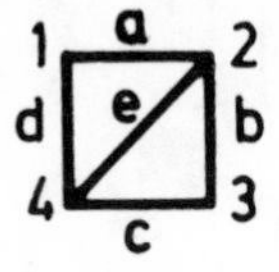

G

cycles:

$c_1 = \{a, e, d\}$

$c_2 = \{b, c, e\}$

$c_3 = \{a, b, c, d\}$

$$\mathbf{C}(G) = \begin{array}{c} \\ (c_1) \\ (c_2) \\ (c_3) \end{array} \begin{array}{c} \begin{array}{ccccc} (a) & (b) & (c) & (d) & (e) \end{array} \\ \begin{bmatrix} 1 & 0 & 0 & 1 & 1 \\ 0 & 1 & 1 & 0 & 1 \\ 1 & 1 & 1 & 1 & 0 \end{bmatrix} \end{array}$$

$$\mathbf{T}(G) = \begin{array}{c} \\ (1) \\ (2) \\ (3) \\ (4) \end{array} \begin{array}{c} \begin{array}{ccccc} (a) & (b) & (c) & (d) & (e) \end{array} \\ \begin{bmatrix} 1 & 0 & 0 & 1 & 0 \\ 1 & 1 & 0 & 0 & 1 \\ 0 & 1 & 1 & 0 & 0 \\ 0 & 0 & 1 & 1 & 1 \end{bmatrix} \end{array}$$

$$\mathbf{T}^{T}(G) = \begin{bmatrix} 1 & 1 & 0 & 0 \\ 0 & 1 & 1 & 0 \\ 0 & 0 & 1 & 1 \\ 1 & 0 & 0 & 1 \\ 0 & 1 & 0 & 1 \end{bmatrix}$$

$$\mathbf{C}(G)\,\mathbf{T}^{T}(G) = \begin{bmatrix} 1 & 0 & 0 & 1 & 1 \\ 0 & 1 & 1 & 0 & 1 \\ 1 & 1 & 1 & 1 & 0 \end{bmatrix} \cdot \begin{bmatrix} 1 & 1 & 0 & 0 \\ 0 & 1 & 1 & 0 \\ 0 & 0 & 1 & 1 \\ 1 & 0 & 0 & 1 \\ 0 & 1 & 0 & 1 \end{bmatrix} = \begin{bmatrix} 2 & 2 & 0 & 2 \\ 0 & 2 & 2 & 2 \\ 2 & 2 & 2 & 2 \end{bmatrix}$$

$$= \mathbf{0}(\text{mod } 2)$$

IV. THE DISTANCE MATRIX

The distance matrix of the graph G, **D**(G), is a real symmetric N × N matrix, which contains elements $D_{ij}(G)$ representing the length of the shortest path (i.e., the minimum number of edges) between *i*-th vertex and *j*-th vertex of G. All diagonal elements $D_{ii}(G)$ are, by definition, zero.

Example

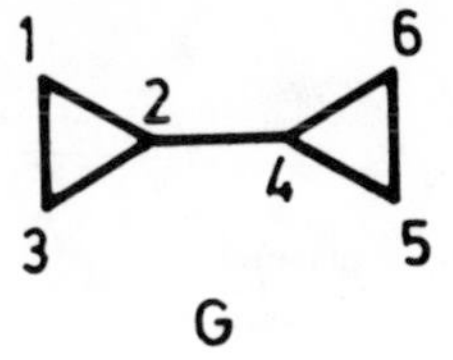

$$\mathbf{D}(G) = \begin{array}{c|cccccc} & ① & ② & ③ & ④ & ⑤ & ⑥ \\ \hline ① & 0 & 1 & 1 & 2 & 3 & 3 \\ ② & 1 & 0 & 1 & 1 & 2 & 2 \\ ③ & 1 & 1 & 0 & 2 & 3 & 3 \\ ④ & 2 & 1 & 2 & 0 & 1 & 1 \\ ⑤ & 3 & 2 & 3 & 1 & 0 & 1 \\ ⑥ & 3 & 2 & 3 & 1 & 1 & 0 \end{array}$$

The distance matrix has found so far only a limited application in chemistry. It was used implicitly by Wiener[16] and Platt[17] in their studies on additive physical parameters of acyclic structures. Wiener has introduced a *path number* W, which is defined as the sum of the distances between any two carbon atoms in alkane in terms of the carbon-carbon bonds. We call the path number the *Wiener number* of the graph G, W(G). It can be shown that the Wiener number is equal to a half-sum of the off-diagonal elements of the distance matrix,

$$W(G) = \frac{1}{2}\sum D_{ij}(G) \tag{47}$$

Example

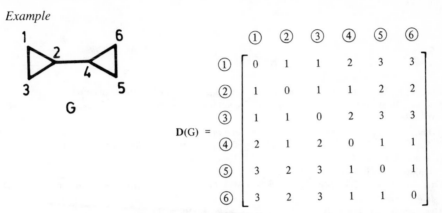

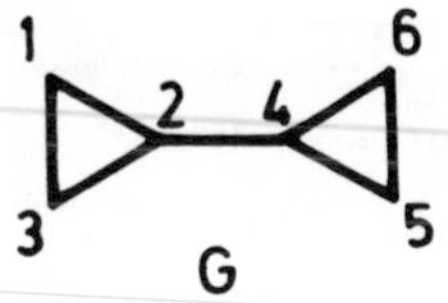

$$W(G) = 27$$

$$\sum_{(i,j)} D_{ij}(G) = 54$$

$$W(G) = \frac{1}{2}\sum_{(i,j)} D_{ij}(G) = \frac{1}{2}\,54 = 27$$

Rouvray[18] has used an index R(G) in his studies of the thermodynamic properties of alkanes, which is the sum of all off-diagonal entries to the distance matrix,

$$R(G) = \sum_{(i,j)} D_{ij} \tag{48}$$

However, the Rouvray index is equal to twice the Wiener number of a graph,

$$R(G) = 2W(G) \tag{49}$$

Clark and Kettle[19] have employed the distance matrix for studying the permutational isomers of stereochemically nonrigid molecules, while Bonchev and Trinajstić[20,21] have used it for characterization of molecular branching. The distance matrix has found some use in other fields, i.e., anthropology.[22]

The distance matrix is related to the adjacency matrix of a graph by the relation,

$$\mathbf{D}(G) = \sum_{i=1}^{N} \mathbf{A}_i(G) = \mathbf{A}(G) + \sum_{i=2} \mathbf{A}_i(G) \tag{50}$$

where $\sum_{i=2}^{N} \mathbf{A}_i(G)$

are topological matrices containing only those nonzero entries which represent the shortest paths between the second, third,......., etc. neighbors.

REFERENCES

1. **Harary, F.,** *Graph Theory,* Addison-Wesley, Reading, Mass., 1971, 150; second printing.
2. **Rouvray, D. H.,** in *Chemical Applications of Graph Theory,* Balaban, A. T., Ed., Academic Press, London, 1977, 175.
3. **Trinajstić, N.,** in *Semiempirical Methods of Electronic Structure Calculations. Part A: Techniques,* Vol. 7, Segal, G. A., Ed., Modern Theoretical Chemistry, Plenum Press, New York, 1977, 1.
4. **Schmidtke, H. H.,** *J. Chem. Phys.,* 45, 3920, 1966.
5. **Schmidtke, H. H.,** *Coord. Chem. Rev.,* 2, 3, 1967.
6. **Graovac, A. and Trinajstić, N.,** *Math. Chem. (Mülheim/Ruhr),* 1, 159, 1975; erratum *Math. Chem. (Mülheim/Ruhr),* 5, 290, 1979.
7. **Graovac, A. and Trinajstić, N.,** *Croat. Chem. Acta,* 47, 95, 1975.
8. **Graovac, A. and Trinajstić, N.,** *J. Mol. Struct.,* 30, 416, 1976.
9. **Randić, M.,** *J. Chem. Phys.,* 60, 3920, 1974.
10. **Ham, N. S.,** *J. Chem. Phys.,* 29, 1229, 1958.
11. **Finkbeiner, D. T., II,** *Introduction to Matrices and Linear Transformations,* W. H. Freeman, San Francisco, 1960.
12. **Kowalewski, G.,** *Determinantentheorie,* 3rd ed., Chelsea, New York, 1943.
13. **Minc, H.,** *Permanents,* Addison-Wesley, Reading, Mass., 1978.
14. **Balandin, A. A.,** *Acta Physicochim. U.R.S.S.,* 12, 447, 1940.
15. **Kirchhoff, G.,** *Ann. Phys. Chem.,* 72, 497, 1847.
16. **Wiener, H.,** *J. Am. Chem. Soc.,* 69, 17, 1947.
17. **Platt, J. R.,** *J. Chem. Phys.,* 15, 419, 1947.
18. **Rouvray, D. H.,** *Math. Chem. (Mülheim/Ruhr),* 1, 125, 1975.
19. **Clark, M. J. and Kettle, S. F. A.,** *Inorg. Chim. Acta,* 14, 201, 1975.
20. **Bonchev, D. and Trinajstić, N.,** *J. Chem. Phys.,* 67, 4517, 1977.
21. **Bonchev, D. and Trinajstić, N.,** *Int. J. Quantum Chem., Symp.,* 12, 293, 1978.
22. **Hage, P.,** *Anthropol. Forum,* 3, 280, 1973.

Chapter 5

THE CHARACTERISTIC POLYNOMIAL OF A GRAPH

I. THE SPECTRUM OF A GRAPH

The adjacency matrix of a graph G, **A**(G), with N vertices could be submitted to various transformations. The linear transformation leading to the *diagonal form,*

$$\mathbf{X} = \begin{bmatrix} x_1 & & & & 0 \\ & x_2 & & & \\ & & \cdot & & \\ & & & \cdot & \\ 0 & & & & x_N \end{bmatrix} \tag{1}$$

is of the greatest importance. Elements $x_1, x_2, \ldots, x_N$ of a diagonal matrix **X** are called the eigenvalues of the matrix **A**(G). Conventionally,

$$x_1 \geqslant x_2 \geqslant \ldots \geqslant x_N \tag{2}$$

If some eigenvalue appears k times, it is k-fold degenerate. The set of all eigenvalues $\{x_1, x_2, \ldots, x_N\}$ is collectively called the *spectrum* of a graph G, i.e., the *graph spectrum.* The graph spectrum is an important graph invariant.[1]

Because the adjacency matrix **A**(G) of a graph is symmetric matrix by definition, the above transformation can be always carried out and furthermore x_j (j = 1,2,.....,N) are all real numbers.[2]

The symmetric matrix is characterized by,

$$\mathbf{A} = \mathbf{A}^T \tag{3}$$

where $\mathbf{A}^T$ is the *transpose* of a matrix **A**. The transpose of a matrix is defined by,

$$A_{ij} = A_{ji} \tag{4}$$

The transformation to a diagonal form is realized by the *eigenvector matrix* **C** of **A**(G),

$$\mathbf{C}\,\mathbf{A}(G) = \mathbf{X}\,\mathbf{C} \tag{5}$$

where **C** is a N × N matrix of the form,

$$\mathbf{C} = \begin{bmatrix} C_1 \\ C_2 \\ \cdot \\ \cdot \\ \cdot \\ C_N \end{bmatrix} \tag{6}$$

The (1 × N) row matrices $\mathbf{C}_i$ (i = 1,2,...,N) are given by,

$$\mathbf{C}_i = [c_{i1}, c_{i2}, \ldots, c_{iN}] \tag{7}$$

Relation (5) may be rewritten, accordingly, as,

$$\mathbf{C}_i\, \mathbf{A}(G) = x_i\, \mathbf{C}_i;\; i = 1,2,\ldots,N \tag{8}$$

where $\mathbf{C}_i$ are eigenvectors belonging to the eigenvalue x_i. Equation (8) may be also presented in the different form,

$$\mathbf{C}_i\, [x_i \mathbf{I} - \mathbf{A}(G)] = 0 \tag{9}$$

where the symbol **I** stands for a unit matrix. The unit matrix **I** is a special case of a diagonal matrix because all its elements are zero except the diagonal elements which are all unity.

Equation (9) represents a system of homogeneous linear equations, that is, equations whose constant terms are zero. They are also called the *secular equations*[3] and are expressed below in a different way,

$$\sum_{k=1}^{N} c_{ik}\, [x_i\, \delta_{k\ell} - A_{k\ell}(G)] = 0;\, i, \ell = 1,2,\ldots,N \tag{10}$$

where $\delta_{k\ell}$ is the Kronecker delta function with values,

$$\delta_{k\ell} = \begin{cases} 1 & k = \ell \\ 0 & k \neq \ell \end{cases} \tag{11}$$

In order that the secular equations (10) have nontrivial solutions, it is necessary that the corresponding *secular determinant* vanishes,[3]

$$\det \left| x\mathbf{I} - A(G) \right| = 0 \tag{12}$$

where the matrix $[x\mathbf{I} - A(G)]$ is the *secular matrix*. The secular matrix is *Hermitian*. A matrix **A** is Hermitian or self-adjoint matrix if,

$$\mathbf{A}^{\dagger} = \mathbf{A} \tag{13}$$

where $\mathbf{A}^{\dagger}$ is the adjoint of a matrix **A**. The adjoint of a given matrix **M** is defined as,

$$\mathbf{M}^{\dagger} = (\mathbf{M}^{*})^{T} \tag{14}$$

where $\mathbf{M}^*$ is the complex conjugate of a matrix **M**. Complex conjugate, M^*, is formed by taking the complex conjugate ($i \rightarrow -i$) of each element, where $i = \sqrt{-1}$. A Hermitian matrix can be converted to a diagonal form by a *similarity transformation*,

$$\mathbf{U}\,\mathbf{A}\,\mathbf{U}^{-1} = \mathbf{X} \tag{15}$$

where **U** is the *unitary matrix*. Because the transforming matrix in the above similarity transformation is unitary, the transformation is referred to as a *unitary transformation*. This procedure is called the *diagonalization* of a matrix. A matrix is unitary, if,

$$\mathbf{U}^{-1} = \mathbf{U}^{\dagger} \tag{16}$$

where $\mathbf{U}^{-1}$ is an inverse of a matrix **U**. The inverse of the matrix **U** is defined such that,

$$\mathbf{U}\,\mathbf{U}^{-1} = \mathbf{U}^{-1}\,\mathbf{U} = \mathbf{I} \tag{17}$$

Thus, the unitary transformation may be presented as,

$$\mathbf{U}\,\mathbf{A}\,\mathbf{U}^{\dagger} = \mathbf{X} \tag{18}$$

The polynomial,

$$P(G; x) = \det \left| xI - A(G) \right| \qquad (19)$$

is called the *characteristic* (or *secular*) *polynomial* of the adjacency matrix of a graph (or in short the characteristic polynomial of a graph G).

P(G; x) is the polynomial of the degree N given by,

$$P(G; x) = \sum_{n=0}^{N} a_n (G) x^{N-n} \qquad (20)$$

where $a_n(G)$ ($n = 0,1,2,\ldots,N$) are the coefficients of the polynomial. The set of all zeros of the characteristic polynomial of a graph forms a *graph spectrum*. P(G; x) can be also given in terms of its solutions,[2]

$$P(G; x) = \prod_{i=1}^{N} (x - x_i) \qquad (21)$$

The interval in which the eigenvalues of a graph lie is limited. According to Frobenius theorem,[4] the limits of the graph spectrum are determined by the maximal degree of a vertex, D_{max}, in a graph,

$$-D_{max} \leqslant x_i \leqslant D_{max} \qquad (22)$$

II. SACHS FORMULA

A lot of the research has been carried out on generating the coefficients $a_n(G)$ of PG; x) from the structure of the graph.[1,4-16] Most elegant, if not the most practical,[11] is the method of Sachs,[17] which can be best summarized in the following formula,[18]

$$a_n(G) = \sum_{s \in S_n} (-1)^{c(s)} 2^{r(s)} \qquad (23)$$

where s is a *Sachs graph*, S_n a set of all Sachs graphs with exactly n vertices, while $c(s)$ and $r(s)$ denote, respectively, the total number of components and the total number of cycles (rings) in s. The components of a Sachs graph can be either complete graphs K_2, i.e., isolated edges, or cycles C_m($m = 3,4,\ldots,N$), or combinations of $\ell \cdot K_2$ and $k \cdot C_m$, with the restriction $2\ell + k \cdot m = n$. Thus, the Sachs graph is a specified subgraph of G. In the mathematical literature such subgraphs are called *mutation graphs,*[13-15] while in the chemical applications of graph theory there is another proposal to name them as the *characteristic graphs.*[19] Sachs called them in his work the *basic figures,* i.e., *"Grundfiguren"*.[17]

The characteristic polynomial may be now expressed in a different form by introducing Equation (23) into (20),

$$P(G; x) = \sum_{n=0}^{N} \sum_{s \in S_n} (-1)^{c(s)} 2^{r(s)} x^{N-n} \qquad (24)$$

This relation can help us to see, in a direct way, how the structure of P(G; x) reflects the topology of a given graph (molecule).

Sachs graphs are related to permutations of nonzero elements $A_{k\ell}$ in the secular determinant. Taking into account K_2 components corresponds to considering only products $A_{k\ell} \cdot A_{\ell k}$ which are related to edges in G. In the case of directed graphs, the product $A_{k\ell} \cdot A_{\ell k}$ really means a cycle between two vertices,

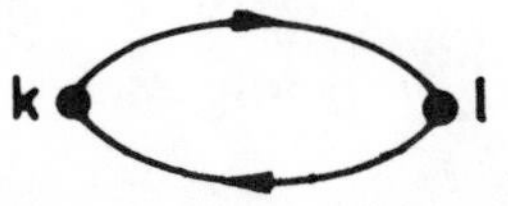

Counting cyclic components C_m ($m = 3,4,\ldots,N$) corresponds to taking into account only products such as $A_{k\ell} \cdot A_{\ell m} \cdot A_{mn} \cdot \ldots \cdot A_{zk}$ and $A_{\ell k} \cdot A_{m\ell} \cdot A_{nm} \cdot \ldots \cdot A_{kz}$, which are products of nonvanishing elements related to cycles in G. There are two products because each cycle may be counted in either of two directions,

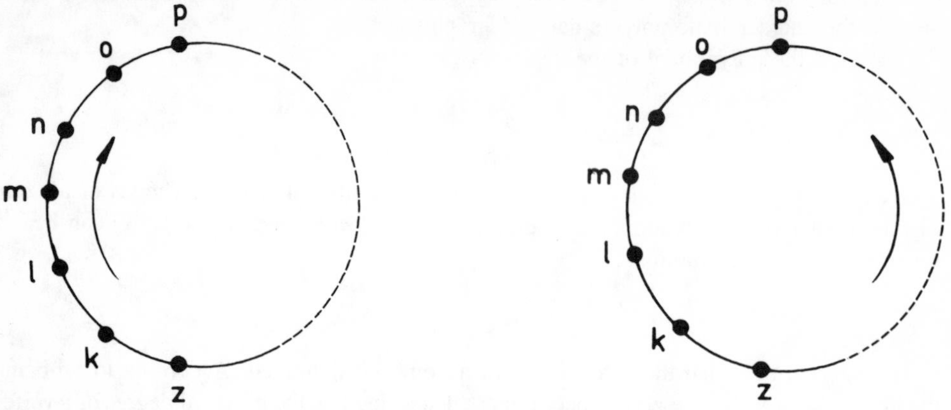

This is the origin of the factor 2 in Equation (23).

III. APPLICATION OF SACHS FORMULA

Let us, for example, consider the graph G corresponding to methylenecyclopropene.

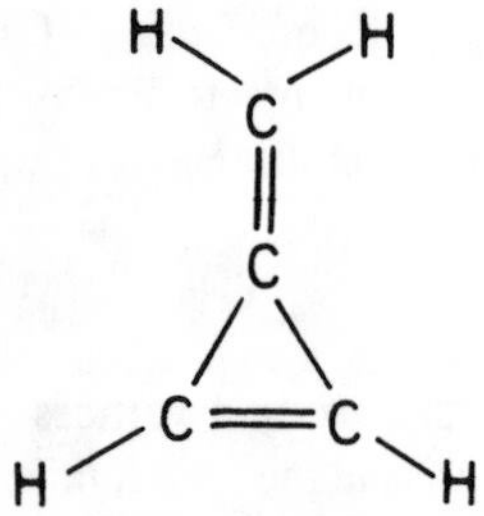

methylenecyclopropene

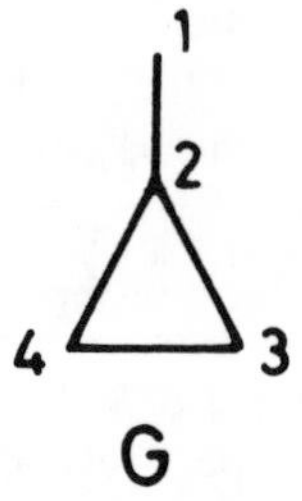

We first set up the adjacency matrix of G. Only nonvanishing entries $A_{k\ell} = 1$ are included in **A**(G),

$$\mathbf{A}(G) = \begin{bmatrix} 0 & A_{12} & 0 & 0 \\ A_{21} & 0 & A_{23} & A_{24} \\ 0 & A_{32} & 0 & A_{34} \\ 0 & A_{42} & A_{43} & 0 \end{bmatrix} \tag{25}$$

Then we produce the corresponding secular determinant,

$$\det \left| xI - \mathbf{A}(G) \right| = \begin{bmatrix} x & -A_{12} & 0 & 0 \\ -A_{21} & x & -A_{23} & -A_{24} \\ 0 & -A_{32} & x & -A_{34} \\ 0 & -A_{42} & -A_{43} & x \end{bmatrix} \tag{26}$$

Table 1
THE CONNECTION BETWEEN THE PRODUCTS OF MATRIX ELEMENTS AND SACHS GRAPHS FOR THE GRAPH DEPICTING METHYLENECYCLOPROPENE

n	Product of matrix elements	Sachs graphs	Numerical value
0			1
1		Ø	0
2	$A_{12}\cdot A_{21} + A_{23}\cdot A_{32} + A_{24}\cdot A_{42} + A_{34}\cdot A_{43}$	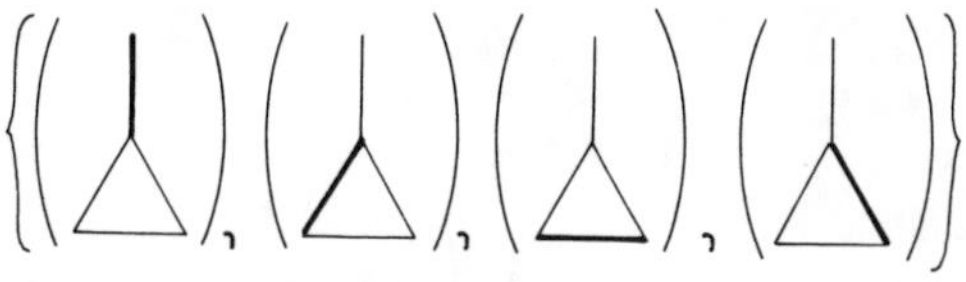	4
3	$A_{23}\cdot A_{34}\cdot A_{42} + A_{32}\cdot A_{43}\cdot A_{24}$	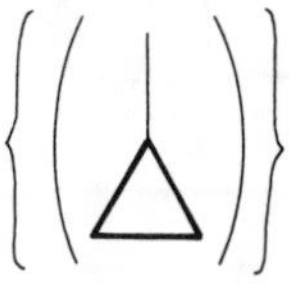	2
4	$A_{12}\cdot A_{21}\cdot A_{34}\cdot A_{43}$	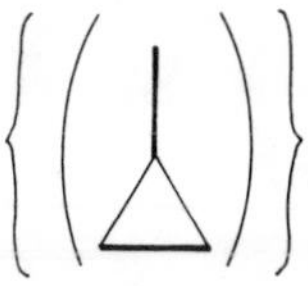	1

The expansion of the above secular determinant generates the characteristic polynomial of G,

$$P(G;x) = x^4 - (A_{12}\cdot A_{21} + A_{23}\cdot A_{32} + A_{24}\cdot A_{42} + A_{34}\cdot A_{43})\,x^2 - (A_{23}\cdot A_{34}\cdot A_{42} + A_{32}\cdot A_{43}\cdot A_{24})\,x + A_{12}\cdot A_{21}\cdot A_{34}\cdot A_{43} \tag{27}$$

Now we wish to investigate the relation between the products of the matrix elements and the matching Sachs graphs. This is accomplished in Table 1.

Substituting numerical values for products of matrix elements back into (27), the explicit form of the characteristic polynomial of G is obtained,

$$P(G;x) = x^4 - 4x^2 - 2x + 1 \tag{28}$$

with the following graph spectrum,

$$\{2.17,\ 0.31,\ -1.00,\ -1.48\} \tag{29}$$

The identical result can be achieved by the *direct* use of the Sachs formula without going through the expansion of the secular determinant. This is given in Table 2.

The way we carried out the construction of P(G; x) for methylenecyclopropene in Table 2 by using the Sachs formula is based on the (combinatorial) partitioning of the molecular graph G into the subgraphs K_2 and/or C_m (m = 3,4,...,N) obeying the restriction $2\ell + k\cdot m = n$.

IV. THE CHARACTERISTIC POLYNOMIAL OF A CHAIN

Linear polyenes can be depicted by chains whose characteristic polynomials are symbolized, for brevity, by L_N.

Table 2
CONSTRUCTION OF THE CHARACTERISTIC POLYNOMIAL OF METHYLENECYCLOPROPENE BY WAY OF THE SACHS FORMULA

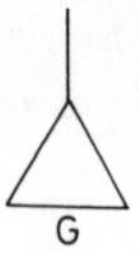

n	S_n		a_n
0			1
1	Ø		0
2	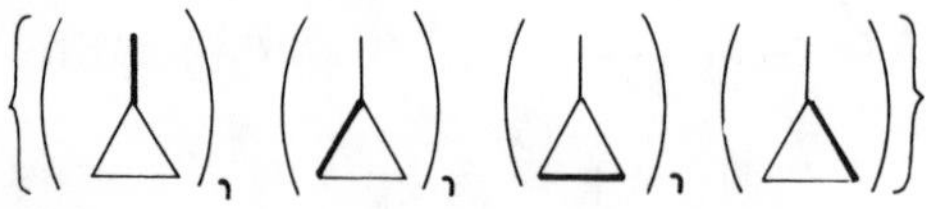	$c(s) = 1$, $r(s) = 0$	$4(-1)\,2^0 = -4$
3	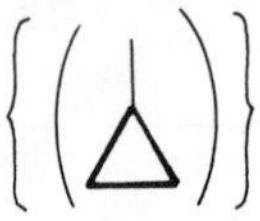	$c(s) = 1$, $r(s) = 1$	$(-1)^1\,2^1 = -2$
4	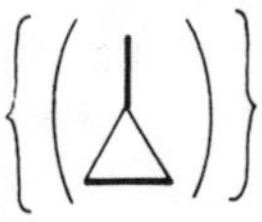	$c(s) = 2$, $r(s) = 0$	$(-1)^2\,2^0 = 1$

$$P(G; x) = x^4 - 4x^2 - 2x + 1$$

Example

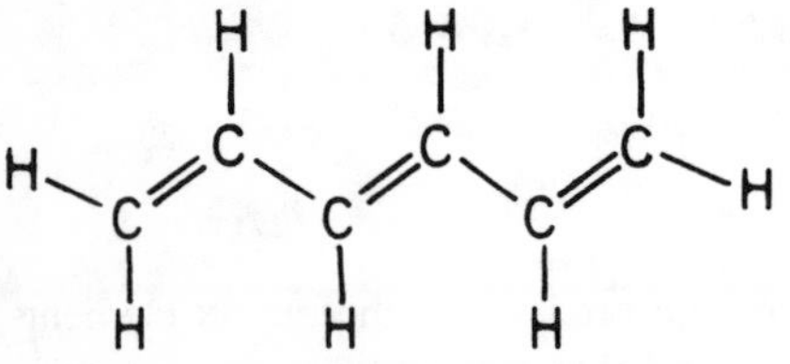

hexatriene

chain with six vertices

L_6

The basic recurrence relation for the graphic evaluation of L_n is given by,[20]

$$L_N = L_{n-e} - L_{N-(e)} \tag{30}$$

where L_{N-e} is obtained by deletion of the end edge *e* from L_N, while $L_{N-(e)}$ is obtained by removal of the edge *e* and both vertices incident to it. Hence, L_{N-e} and $L_{N-(e)}$ possess N and N − 2 vertices, respectively.

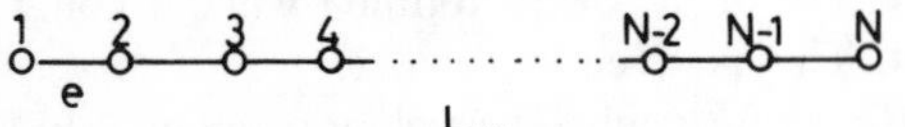

L_N

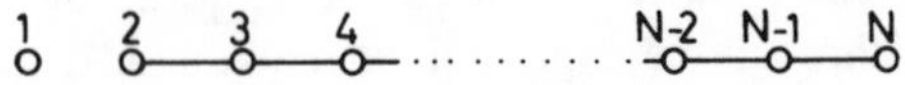

$L_{N-e} \equiv xL_{N-1}$

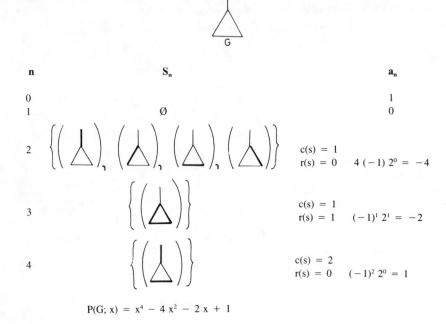

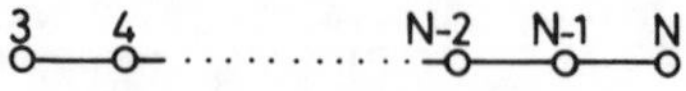

$L_{N-(e)} \equiv L_{N-2}$

The result $L_N = xL_{N-1}$ is a consequence of making the second row and the second column of the determinant of L_N to be zero when the end edge of L_N is removed.

If xL_{N-1} and L_{N-2} are substituted for L_{N-e} and $L_{N-(e)}$ into (30) the recurrence relation is obtained,[21]

$$L_N = x L_{N-1} - L_{N-2} \tag{31}$$

which enables the calculation of the characteristic polynomials of polyenes starting with,

$$L_0 = 1 \tag{32}$$

$$L_1 = x \tag{33}$$

In Table 3 L_N polynomials for polyenes up to N = 20 are tabulated.

V. THE CHARACTERISTIC POLYNOMIAL OF A CYCLE

The graph theoretical representation of the [N]-annulene is the cycle C_N with N vertices. The [N]-annulenes are conjugated cyclic hydrocarbons of the general formula C_NH_N ($N \geq 3$). The characteristic polynomial of an N-cycle $P(C_N; x)$ can be obtained by relating it to the polynomials L_N of [N]-polyene and L_{N-2} of [N−2]-polyene,[20] obtained by dissecting the N-cycle,[21]

$$P(C_N; x) = L_N - L_{N-2} - 2 \tag{34}$$

[N]-polyene is obtained from [N]-annulene by excising the edge *e* from the N-cycle, while [N−2]-polyene is generated by removing the edge *e* and adjacent vertices from C_N,

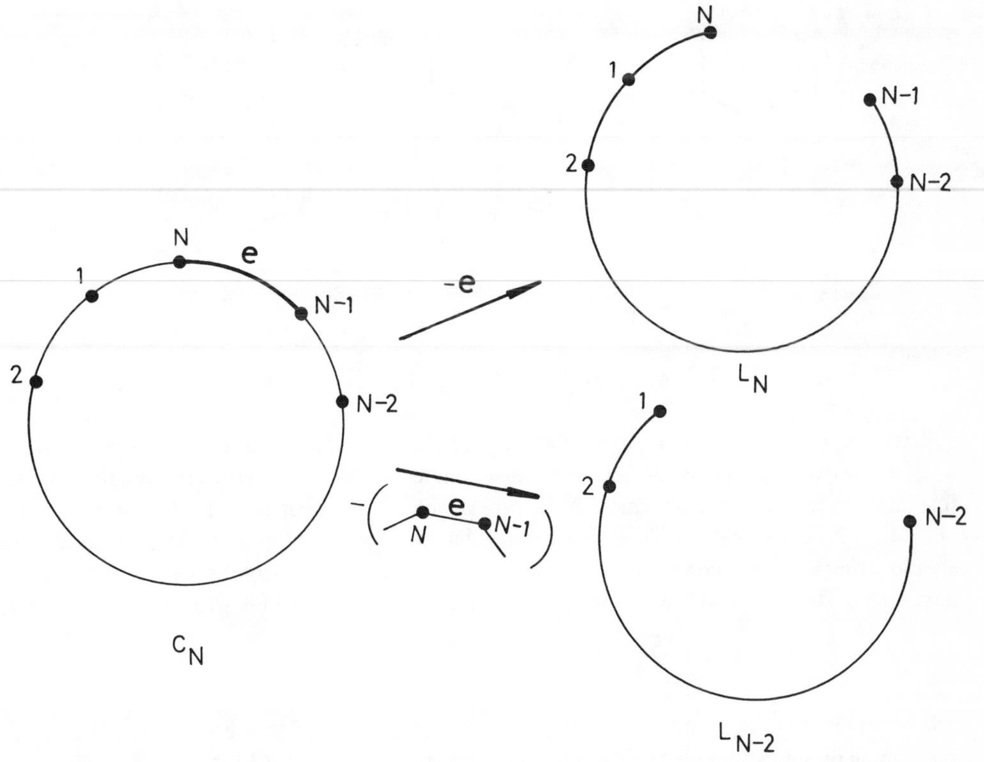

The factor −2 in (34) results from two closed paths in C_N each contributing −1. Since L_N and L_{N-2} are known polynomials one can easily evaluate $P(C_N; x)$ from the relation (34).

Table 3
CHARACTERISTIC POLYNOMIALS OF LINEAR POLYENES

$L_0 = 1$
$L_1 = x$
$L_2 = x^2 - 1$
$L_3 = x^3 - 2x$
$L_4 = x^4 - 3x^2 + 1$
$L_5 = x^5 - 4x^3 + 3x$
$L_6 = x^6 - 5x^4 + 6x^2 - 1$
$L_7 = x^7 - 6x^5 + 10x^3 - 4x$
$L_8 = x^8 - 7x^6 + 15x^4 - 10x^2 + 1$
$L_9 = x^9 - 8x^7 + 21x^5 - 20x^3 + 5x$
$L_{10} = x^{10} - 9x^8 + 28x^6 - 35x^4 + 15x^2 - 1$
$L_{11} = x^{11} - 10x^9 + 36x^7 - 56x^5 + 35x^3 - 6x$
$L_{12} = x^{12} - 11x^{10} + 45x^8 - 84x^6 + 70x^4 - 21x^2 + 1$
$L_{13} = x^{13} - 12x^{11} + 55x^9 - 120x^5 + 126x^5 - 56x^3 + 7x$
$L_{14} = x^{14} - 13x^{12} + 66x^{10} - 165x^8 + 210x^6 - 126x^4 + 28x^2 - 1$
$L_{15} = x^{15} - 14x^{13} + 78x^{11} - 220x^9 + 330x^7 - 252x^5 + 84x^3 - 8x$
$L_{16} = x^{16} - 15x^{14} + 91x^{12} - 286x^{10} + 495x^8 - 462x^6 + 210x^4 - 36x^2 + 1$
$L_{17} = x^{17} - 16x^{15} + 105x^{13} - 364x^{11} + 715x^9 - 792x^7 + 462x^5 - 120x^3 + 9x$
$L_{18} = x^{18} - 17x^{16} + 120x^{14} - 455x^{12} + 1001x^{10} - 1287x^8 + 924x^6 - 330x^4 + 45x^2 - 1$
$L_{19} = x^{19} - 18x^{17} + 136x^{15} - 560x^{13} + 1365x^{11} - 2002x^9 + 1716x^7 - 792x^5 + 165x^3 - 10x$
$L_{20} = x^{20} - 19x^{18} + 153x^{16} - 680x^{14} + 1820x^{12} - 3003x^{10} + 3003x^8 - 1716x^6 + 495x^4 - 55x^2 + 1$

Example

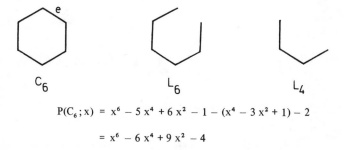

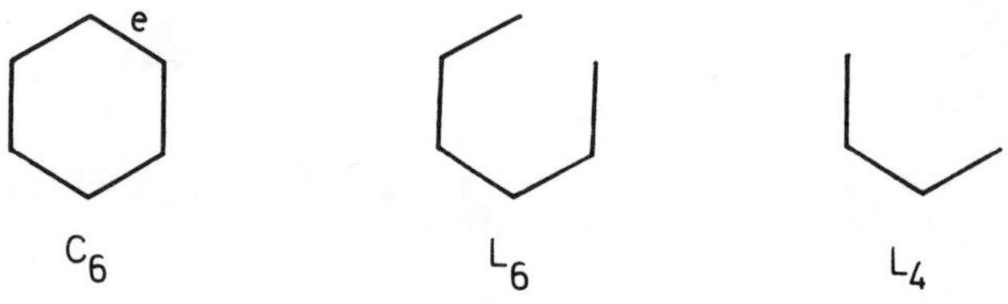

$$P(C_6; x) = x^6 - 5x^4 + 6x^2 - 1 - (x^4 - 3x^2 + 1) - 2$$

$$= x^6 - 6x^4 + 9x^2 - 4$$

The spectrum of C_6 is $\{2, 1, 1, -1, -1, -2\}$.

VI. EXTENSION OF SACHS FORMULA TO MÖBIUS SYSTEMS

The analysis of the nonvanishing elements of the adjacency matrix of the Möbius graph leads to the extended Sachs formula.[22,23] The extended Sachs formula differs from the original formula[18] only in one factor which takes care of the odd number of -1 edges in the C_m ($m = 3,4,\ldots,N$) components. This also means that the Sachs graph s of the Möbius graphs should consist of K_2 components, and/or C_m cycles, and/or Möbius cycles $C_m^{Mö}$ ($m = 3,4,\ldots,N$). The extended Sachs formula which embraces the Möbius structures is given by,

$$a_n(G_{Mö}) = \sum_{s \in S_n} (-1)^{c(s) + p(r)}\, 2^{r(s)} \tag{35}$$

where $p(r)$ is the number of -1 connectivities in the rings of a Sachs graph s. Other symbols have their previous meanings. The characteristic polynomial of a Möbius molecule,

$$P(G_{Mö}; x) = \sum_{n=0}^{N} a_n(G_{Mö})\, x^{N-n} \tag{36}$$

can be directly evaluated using formula (35). As an example Möbius cyclopropenyl, depicted by Möbius graph $G_{Mö}$, is considered,

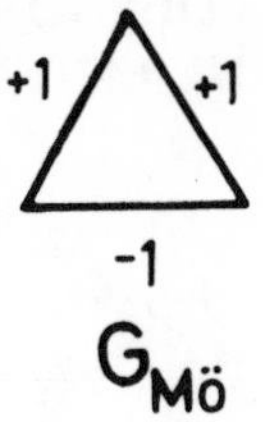

Evaluation of the coefficients a_n ($G_{Mö}$) is reported in Table 4.

The characteristic polynomial of Möbius cyclopropenyl is given by,

$$P(G_{Mö}, x) = x^3 - 3x + 2 \tag{37}$$

with the spectrum: {1, 1, −2}.

Note, that the phase dislocation is meaningless in polyenes,

$$L_N^{Mö} = L_N \tag{38}$$

This is so because in building up the characteristic polynomials of polyenes only Sachs graphs considered are K_2. In the case of "Möbius" polyenes these are related to nonvanishing matrix elements either of the type $A_{k\ell} \cdot A_{\ell k}$ or $(-A_{k\ell}) \cdot (-A_{\ell k})$ with the contribution to the polynomial coefficients in both cases +1.

We remark in passing by that the Möbius graph $G_{Mö}$ is also a special kind of the edge-weighted graph G_{EW} where the edge weight is −1.

VII. THE CHARACTERISTIC POLYNOMIAL OF A MÖBIUS CYCLE

The graph theoretical description of Möbius [N]-annulene is the Möbius cycle $C_N^{Mö}$ with N vertices and at least one (odd) edge with the weight −1. The position of the weighted edge in $C_N^{Mö}$ is between the vertices v_1 and v_N.

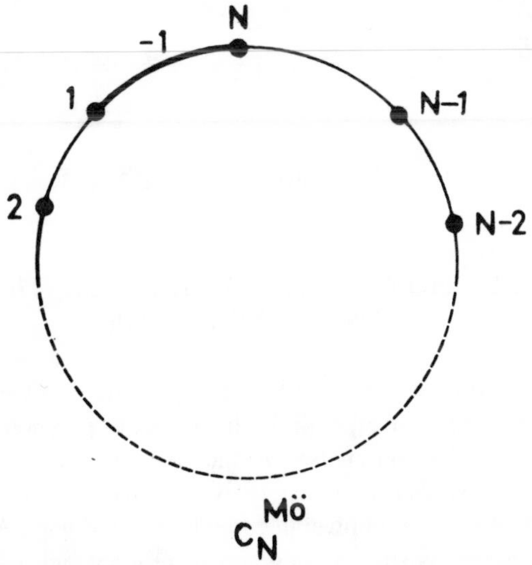

The characteristic polynomial of a Möbius cycle $P(C_N^{Mö}; x)$ can be obtained in the same way as the characteristic polynomial of a cycle $P(C_N; x)$ by dissecting $C_N^{Mö}$ into polyenes, except that the constant factor is now +2,

Table 4

**CONSTRUCTION OF THE COEFFICIENTS OF THE
CHARACTERISTIC POLYNOMIAL FOR MÖBIUS
CYCLOPROPENYL VIA THE EXTENDED SACHS FORMULA**

n	S_n		a_n
0			1
1	Ø		0
2	$\left\{ \left(\triangle \right)_s , \left(\triangle \right)_s , \left(\triangle \right)_s \right\}$	$c(s) = 1$ $r(s) = 0$ $p(r) = 0$	$3\,(-1)^{1\,+\,0}\,2^0 = -3$
3	$\left\{ \left(\underset{-1}{\triangle} \right)_s \right\}$	$c(s) = 1$ $r(s) = 1$ $p(s) = 1$	$(-1)^{1\,+\,1}\,2^1 = 2$

$$P\left(C_N^{M\ddot{o}} ; x \right) = L_N - L_{N-2} + 2 \tag{39}$$

Since L_N and L_{N-2} are polynomials tabulated in Table 3, it is easy to set up the characteristic polynomials for Möbius [N]-annulenes.

Example

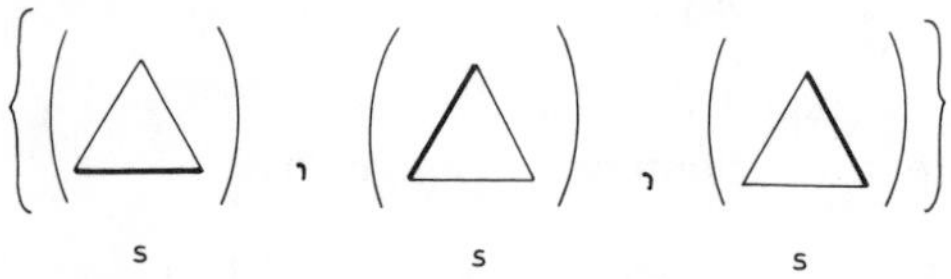

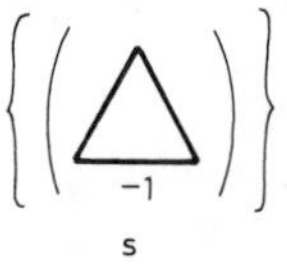

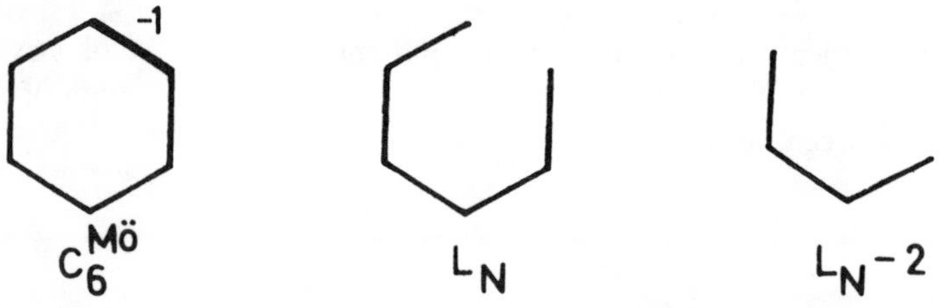

$$P\left(C_6^{M\ddot{o}} ; x \right) = x^6 - 6\,x^4 + 9\,x^2 \text{ with the spectrum: } \left\{ \sqrt{3}, \sqrt{3}, 0, 0, -\sqrt{3}, -\sqrt{3} \right\}.$$

VIII. THE CONSTRUCTION OF THE CHARACTERISTIC POLYNOMIAL OF A WEIGHTED GRAPH

In this section, we wish to demonstrate how the structure of a vertex- and edge-weighted
graph G_{VEW} is related to the corresponding characteristic polynomial $P(G_{VEW}; x)$. Before
doing that we need to introduce a new type of the Sachs graph containing weighted vertices
and/or weighted edges. It is also immediately evident that a graph G_{VEW} may contain both
types of subgraphs, that is to say nonweighted and weighted ones. We define a vertex- and
edge-weighted Sachs graph as such a subgraph of G_{VEW} which has no other components
than one-cycles (loops), and/or (weighted) isolated edges, and/or (weighted) cycles.[24]

A complete set of Sachs graphs of G_{VEW}, corresponding to cyclopropenthione, is given
below,

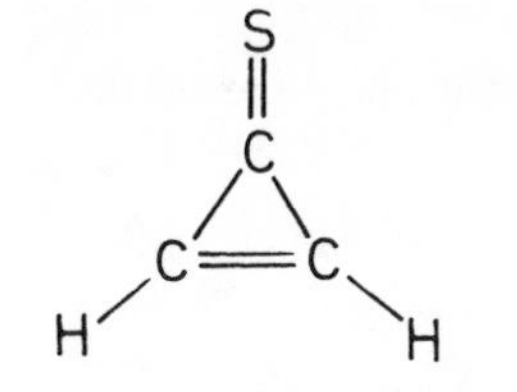

cyclopropenthione

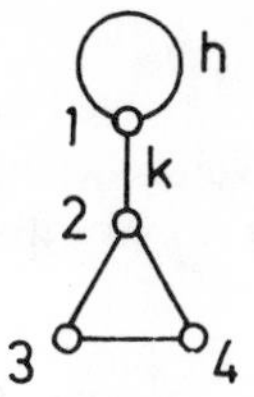

G_{VEW}

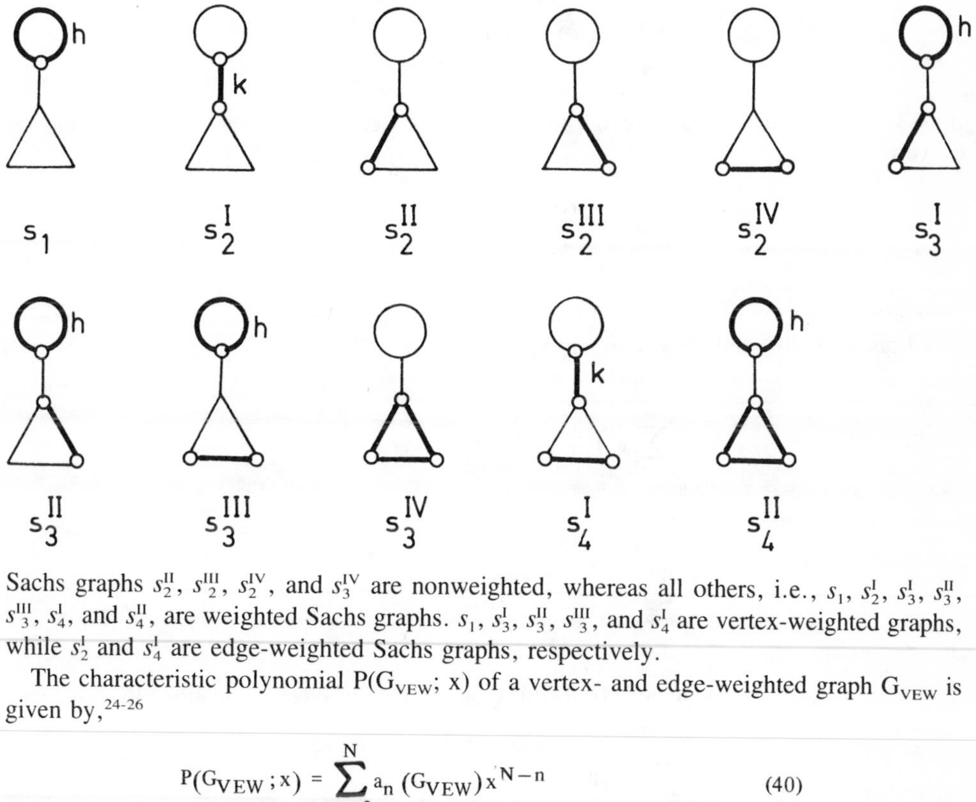

Sachs graphs s_2^{II}, s_2^{III}, s_2^{IV}, and s_3^{IV} are nonweighted, whereas all others, i.e., s_1, s_2^{I}, s_3^{I}, s_3^{II}, s_3^{III}, s_4^{I}, and s_4^{II}, are weighted Sachs graphs. s_1, s_3^{I}, s_3^{II}, s_3^{III}, and s_4^{I} are vertex-weighted graphs, while s_2^{I} and s_4^{I} are edge-weighted Sachs graphs, respectively.

The characteristic polynomial $P(G_{VEW}; x)$ of a vertex- and edge-weighted graph G_{VEW} is given by,[24-26]

$$P(G_{VEW}; x) = \sum_{n=0}^{N} a_n (G_{VEW}) x^{N-n} \tag{40}$$

where

$$a_n (G_{VEW}) = \sum_{s \in S_n} (-1)^{c(s)} 2^{r(s)} \prod_i h_i^{\ell(s)} \prod_j k_j \tag{41}$$

In the above equation, symbols have the following meaning: $\ell(s)$ is the number of loops in a vertex-weighted Sachs graph s, $\prod_i h_i^{\ell(s)}$ gives the contribution from the weighted vertices with the weight h in s, while the index i signifies the number of different weights. In the case when there is only *one* type of weighted vertices, with the weight h, present in G_{VEW}, the product $\prod_i h_i^{\ell(s)}$ reduces to $h^{\ell(s)}$. The product $\prod_j k_j$ reflects the presence of edge-weighted Sachs graphs. Index j indicates the number of differently weighted edges in G_{VEW}. The k-contributions are obtainable in the following way. For a weighted K_2 component of s, k's are squared if they have the same values, otherwise they represent products of two different parameters. This is the consequence of the related matrix elements being $k_i'(-A_{k\ell})\cdot k_i''(-A_{\ell k}) = k_i' \cdot k''_i$. On the other hand if $k_i' = k_i''$, the product is equal to k_i^2. In the case of an edge-weighted cycle C_m, k's are multiplied around the ring. They reflect the product of matrix elements $k_{j1}(-A_{k\ell}) \cdot k_{j2}(-A_{\ell m}) \cdot \ldots \cdot k_{jt}(-A_{zk}) = (-1)^t k_{j1}k_{j2} \ldots k_{jt}$. Of course, if $k_{j1} = k_{j2} = \ldots = k_{jt} = k_j$, then the product is equal to k_j^t.

When the vertex-weighted graph G_{VW} is considered only, the product $\Pi_j k_j$ is equal to unity, and Equations (40) and (41) reduce to,

$$P(G_{VW}; x) = \sum_{n=0}^{N} a_n(G_{VW}) x^{N-n} \quad (42)$$

with

$$a_n(G_{VW}) = \sum_{s \in S_n} (-1)^{c(s)} 2^{r(s)} \prod_i h_i^{\ell(s)} \quad (43)$$

If the vertex-weighted graph contains only one type of weighted vertices, the above equation converts into,

$$a_n(G_{VW}) = \sum_{s \in S_n} (-1)^{c(s)} 2^{r(s)} h^{\ell(s)} \quad (44)$$

Similarly, when the edge-weighted graph G_{EW} is considered only, the product $\Pi_i h_i^{\ell(s)} = 1$, and Equations (40) and (41) contract to,

$$P(G_{EW}; x) = \sum_{n=0}^{N} a_n(G_{EW}) x^{N-n} \quad (45)$$

with

$$a_n(G_{EW}) = \sum_{s \in S_n} (-1)^{c(s)} 2^{r(s)} \prod_j k_j \quad (46)$$

If the edge-weighted graph includes only one type of weighted edges, Equation (46) becomes,

$$a_n(G_{EW}) = \sum_{s \in S_n} (-1)^{c(s)} 2^{r(s)} k^t \quad (47)$$

where t takes values 2 for each disjoint weighted edge and/or the number of weighted edges in the weighted cycle.

Obviously, for the case of $h = 0$ and $k = 1$, Equations (40) and (41) reduce to (20) and (23).

Let us now construct the characteristic polynomial of a vertex- and edge-weighted graph G_{VEW} representing a thiophen-like system by using Equations (40) and (41).

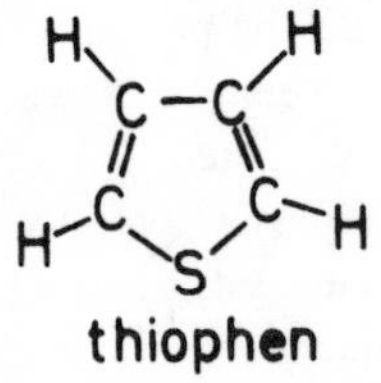

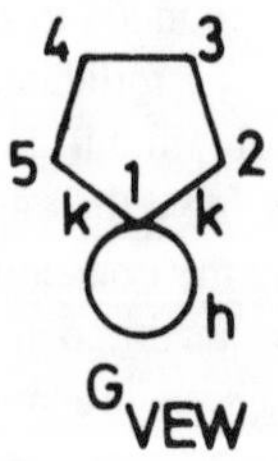

We first give all Sachs graphs of G_{VEW},

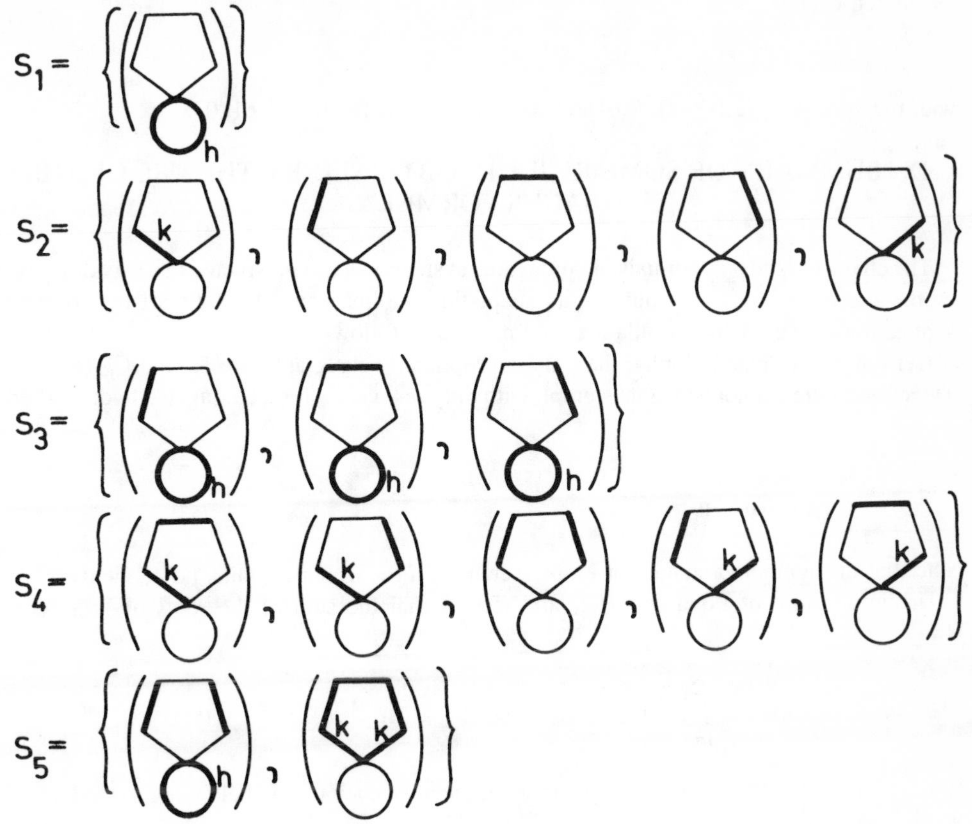

Then the coefficients $a_n(G_{VEW})$ are evaluated using Equation (41):

$$a_0 = 1 \text{ (by definition)}$$

$$a_1 = (-1)^1\, 2^0\, h^1\, k^0 = -h$$

$$a_2 = 2\,(-1)^1\, 2^0\, h^0\, k^2 + 3\,(-1)^1\, 2^0\, h^0\, k^0 = -2k^2 - 3$$

$$a_3 = 3\,(-1)^2\, 2^0\, h^1\, k^0 = 3h$$

$$a_4 = 4\,(-1)^2\, 2^0\, h^0\, k^2 + (-1)^2\, 2^0\, h^0\, k^0 = 4k^2 + 1$$

$$a_5 = (-1)^3\, 2^0\, h^1\, k^0 + (-1)^1\, 2^1\, h^0\, k^2 = -h - 2k^2$$

Finally, the characteristic polynomial of G_{VEW} is constructed,

$$P(G_{VEW}; x) = x^5 - hx^4 - (2k^2 + 3)x^3 + 3hx^2 + (4k^2 + 1)x - (h + 2k^2) \quad (48)$$

Arbitrary selection of parameters $h = 0.5$ and $k = 1.5$, leads to the polynomial,

$$P(G_{VEW}; x) = x^5 - 0.5x^4 - 7.5x^3 + 1.5x^2 + 10x - 5 \quad (49)$$

with the following spectrum,

$$\{2.67110, 0.90266, 0.61803, -1.81803, -2.07375\} \quad (50)$$

Polynomial (48) for $h = 0$ and $k = 1$ becomes the characteristic polynomial of a five-membered cycle,

$$P(C_5, x) = x^5 - 5x^3 + 5x - 2 \quad (51)$$

with the spectrum: {2.00000, 0.61803, 0.61803, −1.61803, −1.61803}.

IX. SUMMARY OF SOME RESULTS OBTAINED BY THE USE OF THE SACHS FORMULA

The characteristic polynomials of conjugated systems can be constructed in a way described by the use of the Sachs formula or its adaptations. Some general results arising from the application of the Sachs formula are summarized as follows:

(A) For nonweighted graphs, the only appropriate Sachs graphs are K_2 and C_m ($m \geq 3$). Therefore, there cannot be a Sachs graph with only *one* vertex for a conjugated hydrocarbon,

$$S_1 = \phi \quad (52)$$

and

$$a_1(G) = 0 \quad (53)$$

Note that the symbol Ø stands for empty set, i.e., a set which contains no elements.

The implication of Equations (52) and (53) is that the sum of the roots of P(G; x) is always zero,

$$a_1(G) = \sum_{i=1}^{N} x_i = 0 \quad (54)$$

which is, in fact, expected because the trace of the adjacency matrix belonging to the Hückel graph is zero.

The nonvanishing value of a_1 coefficients appear only when the vertex- (and edge)-weighted graphs are considered. Thus,

$$a_1(G_{VW}) = a_1(G_{VEW}) = \sum_{i} h_i \quad (55)$$

where h_i stands for the "weight" of the i-th vertex in G_{VW} or G_{VEW}. In addition, the sum of the whole spectrum of G_{VEW} is equal to the sum of the selected parameters for the "weighted" vertices appearing in it,

$$\sum_{i=1}^{N} x_i = \sum_{i} h_i \quad (56)$$

(B) The construction of the characteristic polynomial, via the formula of Sachs, reveals that the a_2 and a_3 coefficients are related to the number of edges (bonds) e and the three-membered rings C_3, respectively, in G,

$$a_2(G) = -e \quad (57)$$

and

$$a_3(G) = -2C_3 \quad (58)$$

In addition, the sum of squares of polynomial roots is related to the a_2 coefficient and consequently, to the number of edges (bonds) in a graph (structure),

$$a_2(G) = -\frac{1}{2}\sum_{i=1}^{N} x_i^2 = -e \tag{59}$$

Relation (59) may be obtained in a simple way by using the result (54),

$$a_2(G) = \sum_{i>j}^{N} x_i x_j$$

$$= \frac{1}{2}\left(\sum_{i=1}^{N}\sum_{j=1}^{N} x_i x_j - \sum_{i=1}^{N} x_i^2\right)$$

$$= \frac{1}{2}\sum_{i=1}^{N} x_i \sum_{j=1}^{N} x_j - \frac{1}{2}\sum_{i=1}^{N} x_i^2$$

$$= -\frac{1}{2}\sum_{i=1}^{N} x_i^2 \tag{60}$$

(C) General expressions for the higher coefficients are rather clumsy. We illustrate this point by presenting expressions for the coefficients a_4, a_5, and a_6 of the characteristic polynomials belonging to Hückel graphs,

$$a_4(G) = \sum_i (2\,K_2)_i - 2\sum_j (C_4)_j \tag{61}$$

$$a_5(G) = \sum_i (K_2 + C_3)_i - 2\sum_j (C_5)_j \tag{62}$$

$$a_6(G) = -\sum_i (3\,K_2)_i + 2\sum_j (K_2 + C_4)_j + 4\sum_k (2\,C_3)_k$$

$$-2\sum_{\ell} (C_6)_{\ell} \tag{63}$$

Each graphical combination within parentheses contributes *one* to the numerical value of the coefficient concerned.

There are more results available along these lines. These will be given later on in the text in a connection with other topics discussed, i.e., the pairing theorem, the number of Kekulé structures, etc.

REFERENCES

1. **Cvetković, D. M., Doob, M., and Sachs, H.,** *Spectra of Graphs,* Academic Press, New York, 1980.
2. **Kurosh, A. G.,** *Higher Algebra,* Mir, Moscow, 1980, third printing.
3. **Coulson, C. A., O'Leary, B., and Mallion, R. B.,** *Hückel Theory for Organic Chemists,* Academic Press, London, 1978, 16.
4. **Coulson, C. A.,** *Proc. Cambridge Phil. Soc.,* 46, 202, 1950.
5. **Samuel, I.,** *C. R. Acad. Sci.,* 229, 1236, 1949.
6. **Gouarné, R.,** *J. Rech. C. N. R. S.,* 34, 81, 1950.
7. **Collatz, L. and Sinogowitz,** *U., Abh. Math. Semin. Univ. Hamburg,* 21, 63, 1957.

8. **Harary, F.,** *SIAM (Soc. Ind. Appl. Math.) Rev.,* 4, 202, 1962.
9. **Spialter, L.,** *J. Am. Chem. Soc.,* 85, 212, 1963.
10. **Spialter, L.,** *J. Chem. Doc.,* 4, 261, 269, 1964.
11. **Hosoya, H.,** *Theor. Chim. Acta,* 25, 215, 1972.
12. **Mowshowitz, A.,** *J. Comb. Theory,* (B)17, 177, 1972.
13. **Schwenk, A. J.,** Ph.D. thesis, The University of Michigan, Ann Arbor, 1973.
14. **Schwenk, A. J.,** in *New Directions in the Theory of Graphs,* Harary, F., Ed., Academic Press, New York, 1973, 275.
15. **Schwenk, A. J.,** in *Graphs and Combinatorics,* Bari, R. and Harary, F., Eds., Springer-Verlag, Berlin, 1974, 153.
16. **Kiang, Y.-S.,** *Int. J. Quantum Chem.,* Symp. 15, 293, 1981.
17. **Sachs, H.,** *Publ. Math. (Debrecen),* 11, 119, 1964.
18. **Graovac, A., Gutman, I., Trinajstić, N., and Živković, T.,** *Theor. Chim. Acta,* 26, 67, 1972.
19. **Randić, M.,** *The Nature of Chemical Structure,* John Wiley & Sons, New York, in press.
20. **Gutman, I., Milun, M., and Trinajstić, N.,** *Croat. Chem. Acta,* 48, 87, 1976.
21. **Heilbronner, E.,** *Helv. Chim. Acta,* 36, 170, 1953.
22. **Graovac, A. and Trinajstić, N.,** *Croat. Chem. Acta,* 47, 95, 1975.
23. **Graovac, A. and Trinajstić, N.,** *J. Mol. Struct.,* 30, 316, 1976.
24. **Trinajstić, N.,** *Croat. Chem. Acta,* 49, 593, 1977.
25. **Graovac, A., Polansky, O. E., Trinajstić, N., and Tyutyulkov, N.,** *Z. Naturforsch.,* 30a, 1696, 1975.
26. **Rigby, M. J., Mallion, R. B., and Day, A. C.,** *Chem. Phys. Lett.,* 51, 178, 1977; erratum *Chem. Phys. Lett.,* 53, 418, 1978.

Chapter 6

TOPOLOGICAL ASPECTS OF HÜCKEL THEORY

Hückel theory is the first introduced, and the simplest, form of the molecular orbital theory of conjugated molecules.[1-3] Since its inception, Hückel theory has been rather successfully used, on a qualitative level, as a guide for chemists in planning and interpreting experiments. Attractive features of the Hückel theory for experimental chemists are its simplicity and limited computational efforts. The last decade has brought forward a number of results which indicate that the success of Hückel theory is based on the fact that it contains intrinsically information about the internal connectivity in the conjugated systems, i.e., it reflects the neighborship of the atoms in conjugated structures.[4,5] In this chapter we will be concerned with the connection between the Hückel theory and topology of the molecular π-network.

I. ELEMENTS OF HÜCKEL THEORY[5-8]

Hückel theory considers only the π-electrons explicitly. This is the Hückel approximation of σ-π separability,

$$<\sigma\,|\,\pi> \;=\; 0 \tag{1}$$

The Hückel molecular orbitals, HMOs, are eigenfunctions ψ_i of the effective one-electron Hamiltonian $H^{\text{Hückel}}$ the precise nature of which is not specified,

$$\hat{H}^{\text{Hückel}}\,\psi_i \;=\; E_i\,\psi_i \tag{2}$$

The quantity E_i is the energy eigenvalue associated with ψ_i. The individual Hückel orbital is expressed as a linear combination of atomic orbitals, *LCAO approximation,* of the form,

$$\psi_i \;=\; \sum_{r=1}^{N} c_{ir}\,\phi_r \tag{3}$$

where ϕ_r is a $2p_z$ orbital on atom r, while c_{ir} is the contribution of the r-th atomic orbital in the i-th molecular orbital. The summation in (3) is over all atoms r in a conjugated molecule.

The total π-electron energy, E_π, is given by,

$$E_\pi \;=\; \sum_{i=1}^{N} g_i\,E_i \;=\; \sum_{i=1}^{N} g_i\,\frac{\sum_r\sum_s c_{ir}^{*}\,H_{rs}\,c_{is}}{\sum_r\sum_s c_{ir}^{*}\,c_{is}\,S_{rs}} \tag{4}$$

where g_i is the occupation number of ψ_i, i.e., the number of electrons in ψ_i, while H_{rs} and S_{rs} are short-hand notations for integrals,

$$H_{rs} \;=\; <r\,|\,\hat{H}^{\text{Hückel}}\,|\,s> \tag{5}$$

$$S_{rs} \;=\; <r\,|\,s> \tag{6}$$

S_{rs} is the overlap integral between π atomic orbitals centered on atoms r and s.

Minimization of E_π by means of the variational method leads to a set of simultaneous, linear, homogeneous equations,

$$\sum_{t=1}^{N} c_{it}\,(H_{rt} - E_i\,S_{rt}) = 0;\; i,r \;=\; 1,2,\ldots,N \tag{7}$$

These equations have nontrivial solutions only if the corresponding Hückel (secular) determinant vanishes,

$$\det | H_{rt} - E_i S_{rt} | = 0 \tag{8}$$

The Hückel determinant can be simplified by use of the Bloch-Hückel approximations,[1,9]

$$\text{(i)}\quad H_{rr} = \langle r | \hat{H}^{\text{Hückel}} | r \rangle = \alpha \tag{9}$$

where H_{rr} is the *Coulomb integral* with the empirical value α. It is assumed that αs are constant for all orbitals ϕ_r centered on similar atoms *r* regardless the variations in the neighboring atoms and groups.

$$\text{(ii)}\quad H_{rt} = \langle r | \hat{H}^{\text{Hückel}} | t \rangle = \begin{cases} \beta & \text{if atoms } r \text{ and } t \text{ are bonded} \\ 0 & \text{otherwise} \end{cases} \tag{10}$$

where H_{rt} is *the resonance integral* with the empirical value β. It is assumed that βs have the same value for π-bonds between the same kind of atoms.

$$\text{(iii)}\quad S_{rt} = \langle r | t \rangle = \delta_{rt} \tag{11}$$

This approximation is rather drastic because it says that there is no overlap between the atoms making up the π-network of a conjugated molecule. However, the neglect of the overlap is justified empirically by the success of the simple HMO procedure in the past 50 years. The inclusion of the overlap among the bonded atoms changes the spacing of energy levels and the values of total π-electron energies, but other quantities are unaffected.[10]

II. EQUIVALENCE BETWEEN HÜCKEL THEORY AND GRAPH SPECTRAL THEORY

Relation (8) may be also given in the matrix form,

$$\det | \mathbf{H} - \mathbf{E}\,\mathbf{S} | = 0 \tag{12}$$

where *H* and *S* are the Hamiltonian and overlap matrices, respectively. As a consequence of the Bloch-Hückel approximations (9) to (11), the matrices *H* and *S* have the following structure,[11]

$$\mathbf{H} = \alpha\mathbf{I} + \beta\mathbf{A} \tag{13}$$

$$\mathbf{S} = \mathbf{I} \tag{14}$$

where **A** is the adjacency matrix of the Hückel graph (conjugated molecule).

The matrix [**H** - **E S**] is called the Hückel matrix. Below we give the Hückel matrix for the π-electrons in fulvene.

Example

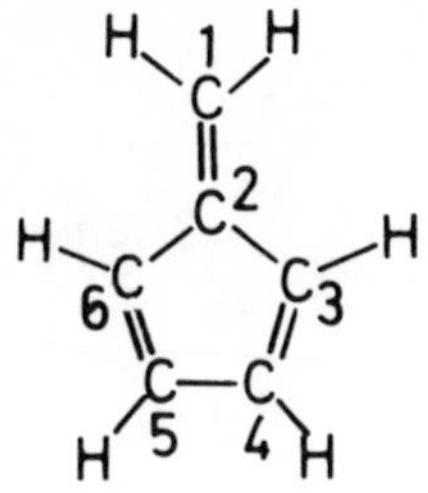

fulvene

$$[\mathbf{H} - \mathbf{ES}] = \begin{array}{c} \\ ① \\ ② \\ ③ \\ ④ \\ ⑤ \\ ⑥ \end{array} \begin{array}{c} \begin{array}{cccccc} ① & ② & ③ & ④ & ⑤ & ⑥ \end{array} \\ \begin{bmatrix} \alpha\text{-E} & \beta & 0 & 0 & 0 & 0 \\ \beta & \alpha\text{-E} & \beta & 0 & 0 & \beta \\ 0 & \beta & \alpha\text{-E} & \beta & 0 & 0 \\ 0 & 0 & \beta & \alpha\text{-E} & \beta & 0 \\ 0 & 0 & 0 & \beta & \alpha\text{-E} & \beta \\ 0 & \beta & 0 & 0 & \beta & \alpha\text{-E} \end{bmatrix} \end{array}$$

Substitution of **H** and **S** by (13) and (14) in (12) and dividing each row by β leads to,

$$\det \left| \frac{E_i - \alpha}{\beta} \mathbf{I} - \mathbf{A} \right| = 0 \qquad (15)$$

Comparison between the determinants (15) and (Equation 12, Chapter 5) indicates that the numbers $E_i - \alpha/\beta$, representing the π energies of individual Hückel orbitals, really constitute the spectrum of eigenvalues of a given Hückel graph,[12]

$$E_i = \alpha + x_i \beta \ ; \ i = 1,2,\ldots,N \qquad (16)$$

If the normalized form of Hückel theory is used, i.e., if β is taken as the energy unit and α as the zero-energy reference point, $\beta = 1$ and $\alpha = 0$, then Equation (16) becomes,

$$E_i = x_i \ ; \ i = 1,2,\ldots,N \qquad (17)$$

The meaning of (17) is that the eigenvalues of the adjacency matrix (i.e., the graph spectrum) are identical with the Hückel orbital energy levels. Since matrices **H** and **A** commute,

$$[\mathbf{H}, \mathbf{A}] = 0 \qquad (18)$$

they have the same set of eigenvectors. Therefore, the eigenvectors of the adjacency matrix are identical with the Hückel molecular orbitals. Hence, Hückel orbitals are sometimes called the *topological orbitals*. From these considerations it is clear that the spectrum of the graph is a rather important quantity in the Hückel-type calculations. The Hückel theory is, in fact, fully equivalent to the graph spectral theory.[12-15] However, the reason for this equivalence is related to the particular nature of the Hückel Hamiltonian; the short-range forces being dominant in the effective potential.[16] Relation (13) shows that the Hückel Hamiltonian is a function of the adjacency matrix,

$$\mathbf{H} = \mathbf{H}(\mathbf{A}) \qquad (19)$$

This leads to the following important conclusion: *the topology of a molecule, rather than its geometry, determines the form of the Hückel molecular orbitals.*

The total π-electron energy (the Hückel energy) of a conjugated molecule in the ground state is given by,

$$E_\pi = E(\text{HMO}) = \sum_{i=1}^{N} g_i E_i \qquad (20)$$

or, by virtue of Equation (17)

$$E_\pi = E(HMO) = \sum_{i=1}^{N} g_i x_i \tag{21}$$

where g_i is the occupation number of i-th orbital which can take values 0, 1, or 2, respectively. The x_i values are conventionally ordered as,

$$x_1 \geqslant x_2 \geqslant \ldots . \geqslant x_N \tag{22}$$

Let the conjugated molecule have N atoms and N_e π-electrons. Then the ground state Hückel energy is given by,

$$E_\pi = \begin{cases} 2\sum_{i=1}^{N_e/2} x_i & \text{if } N_e = \text{even} \\ 2\left[\sum_{i=1}^{(N_e-1)/2} x_i + x_{(N_e+1)/2}\right] & \text{if } N_e = \text{odd} \end{cases} \tag{23}$$

Since for the majority of conjugated systems $N_e = N$, it follows,

$$E_\pi = \begin{cases} 2\sum_{i=1}^{N/2} x_i & \text{if } N = \text{even} \\ 2\left[\sum_{i=1}^{(N-1)/2} x_i + x_{(N+1)/2}\right] & \text{if } N = \text{odd} \end{cases} \tag{24}$$

III. THE HÜCKEL SPECTRUM

The Hückel spectrum is given by an ordered sequence of quantities,

$$\{x_1, x_2, \ldots, x_N\} \tag{25}$$

The extreme values of these quantities are defined by the Frobenius theorem (see discussion in Chapter 5, Section I). Since the maximal topological valency in Hückel graphs is,

$$D_{max} = 3 \tag{26}$$

the interval in which lie all the Hückel eigenvalues is immediately determined,

$$-3 \leqq x_i \leqq +3 \tag{27}$$

Similarly, in the linear polyenes and annulenes $D_{max} = 2$, so the extreme values of Hückel levels for these molecules are ± 2. Finally, since in K_2 graph $D_{max} = D = 1$, the ethylene spectrum consists only of two integral numbers $\{1, -1\}$. There are only five other molecules whose spectra have only integers.[17] These are given in Figure 1.

The set of numbers (25) can be partitioned in three subsets corresponding to the bonding,

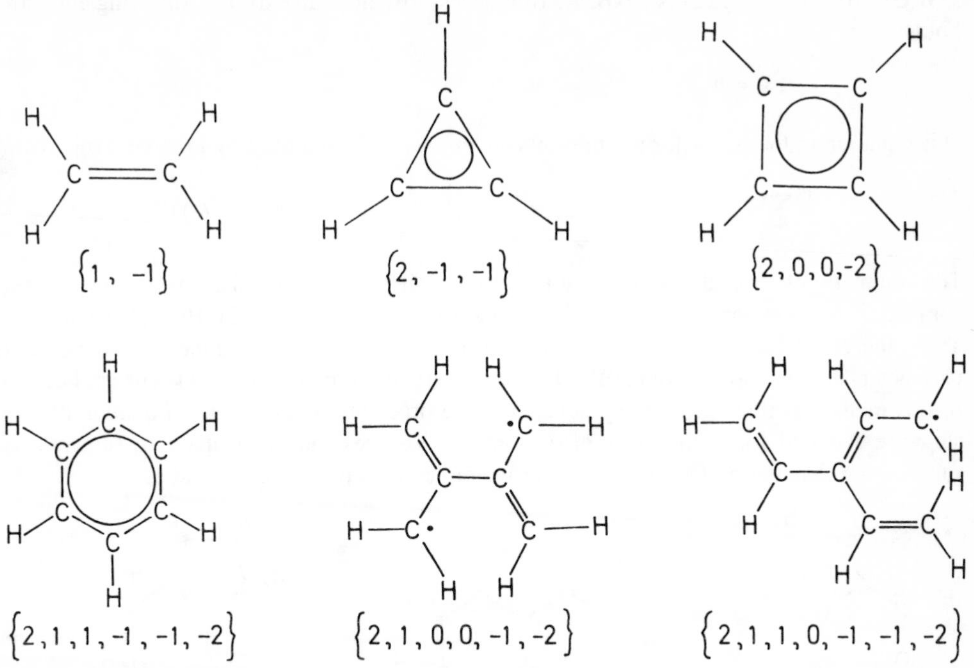

FIGURE 1. The only conjugated systems with integral Hückel spectra.

nonbonding, and antibonding energy levels denoted by N_+, N_0, and N_-, respectively. These are related to the number of atoms N in the conjugated system,

$$N_+ + N_0 + N_- = N \tag{28}$$

These values are important for the chemistry of conjugated molecules. It is especially important to establish whether the nonbonding molecular orbitals, NBMOs, are present in the Hückel spectrum because their existence leads to the prediction[18] that such molecules should have open-shell ground states and be very reactive. Although in reality the situation is much more complicated, for example, because of the Jahn-Teller effects in the case of the triplet ground states,[19] it is an established fact[20] that the structures possessing NBMOs are rarely encountered in the chemistry of conjugated hydrocarbons, and even then these are obtained under the drastic conditions of low temperature chemistry.[21]

A. A Method for the Enumeration of NBMOs

There is a simple way to determine whether a conjugated system has or does not have NBMOs. Since the number of NBMOs is identical with N_0, i.e., the number of zeros in the Hückel spectrum, the determinant of the adjacency matrix vanishes,

$$\det \mathbf{A} = \prod_{i=1}^{N} x_i = 0 \text{ if } x_m = 0 \tag{29}$$

Therefore, the determinant of **A** will be zero if, and only if, there exists at least one zero element in the Hückel spectrum of a molecule. Hence, the problem of whether a conjugated system has or does not have NBMOs can be solved. The question is how to obtain the number of NBMOs without going through the procedure of diagonalization of the Hückel matrix. One way will be described below.

If $\mathbf{C} = (c_1, c_2, \ldots, c_N)$ is a NBMO (not necessarily normalized), the following equation holds,

$$\mathbf{C}\,\mathbf{A} = \mathbf{0} \tag{30}$$

This equation in the scalar form represents *a zero-sum rule* first used by Longuet-Higgins,[18]

$$\sum_{r \to s} c_r = 0; \; s = 1,2,\ldots,N \tag{31}$$

The summation is over all vertices *r* joined to the vertex *s*. The number of the independent parameters in an unnormalized NBMO is equal to the number N_0 in the Hückel spectrum.[22] Thus, the enumeration of N_0 is reduced to a determination of the number of independent parameters in the unnormalized NBMO which satisfy the zero-sum rule (31). The application of this method is illustrated for pyracyclene. The procedure is as follows: Equation (31) is stepwise satisfied for each vertex of a graph G. Vertices for which the zero-sum rule is fulfilled are denoted by ● . We start, for example, with parameters *a* and *b*.

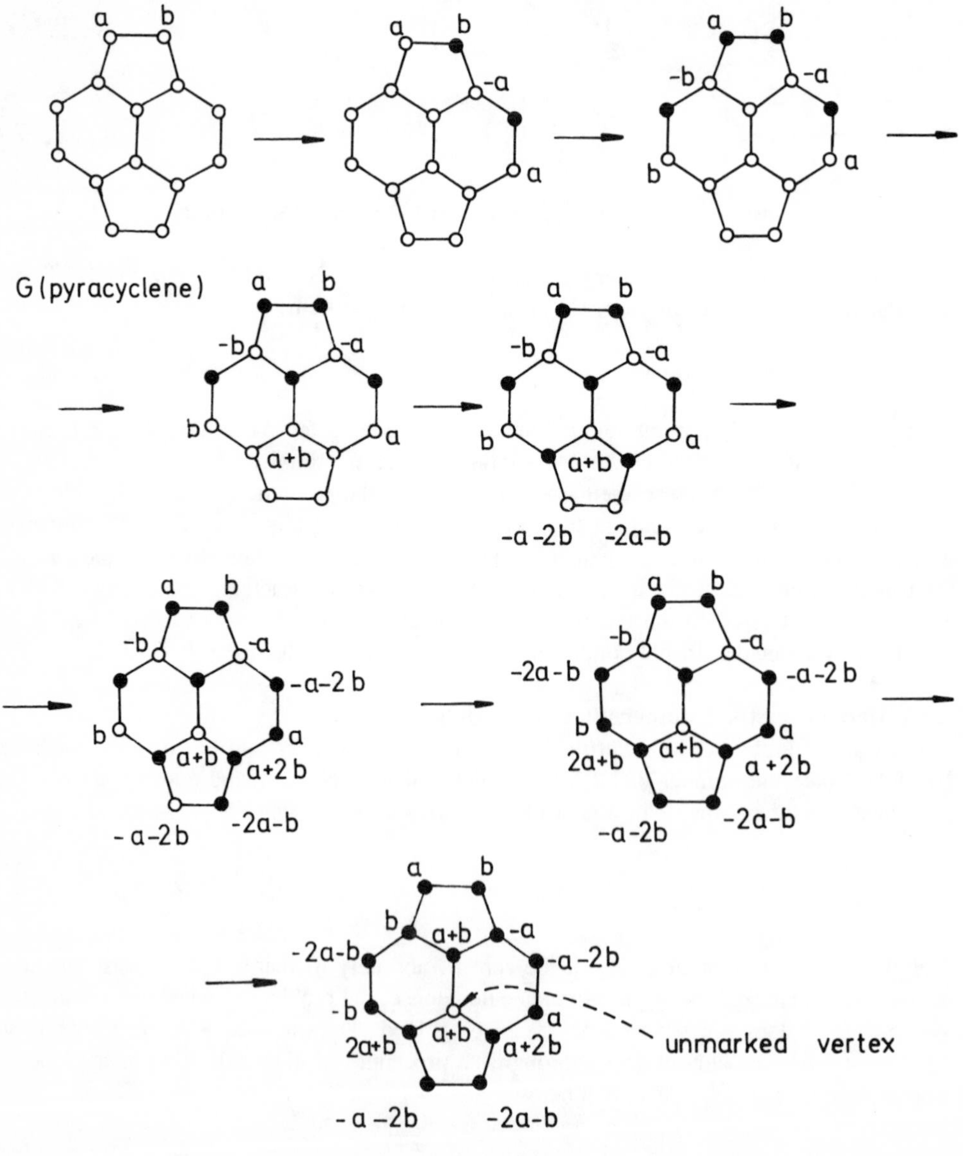

In order that Equation (31) holds for the last (unmarked) vertex of the pyracyclene graph, the following relation must be equal to zero,

$$a + b = 0 \quad (32)$$

If this is so, only *one* parameter is independent,

$$a = -b \quad (33)$$

and consequently

$$N_0(\text{pyracyclene}) = 1 \quad (34)$$

The normalized NBMO of pyracyclene is given below,

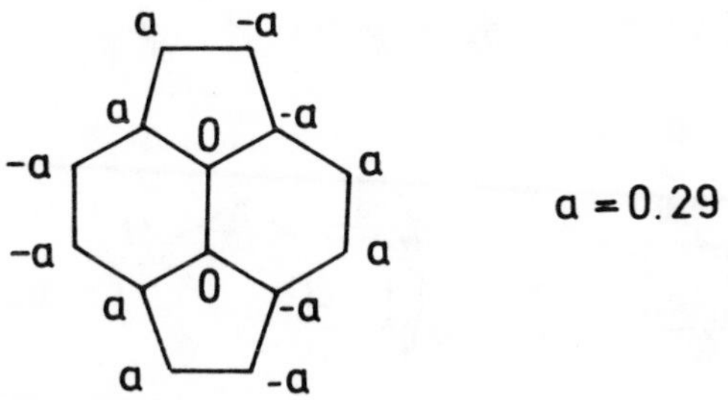

The whole procedure may be simplified by applying certain graph transformations under which the value of N_0 remains the same, but which make this graphical method much easier to apply.[23]

B. Graph-Theoretical Transformations which do not Alter the Value of N_0

In this section, we present several graph-theoretical transformations which do not affect the value of N_0. These are

1. If there is a vertex of degree one, we can remove this vertex, its neighbor, and all adjacent edges without changing the value of N_0,

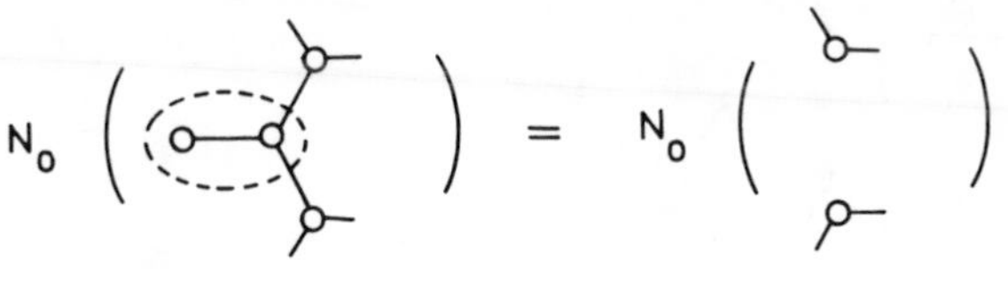

(35)

Example

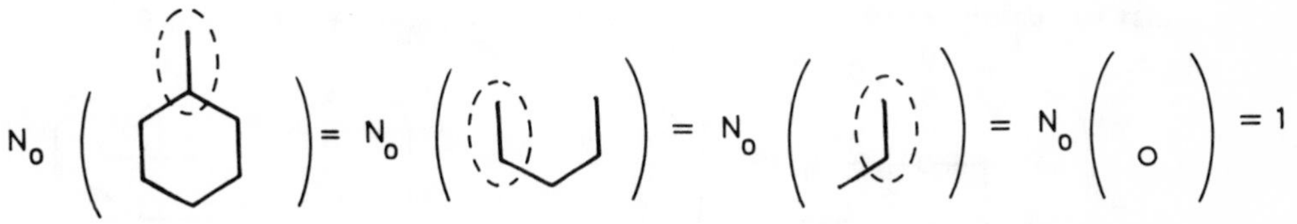

2. A chain of four vertices can be replaced by an edge without changing the value of N_0,

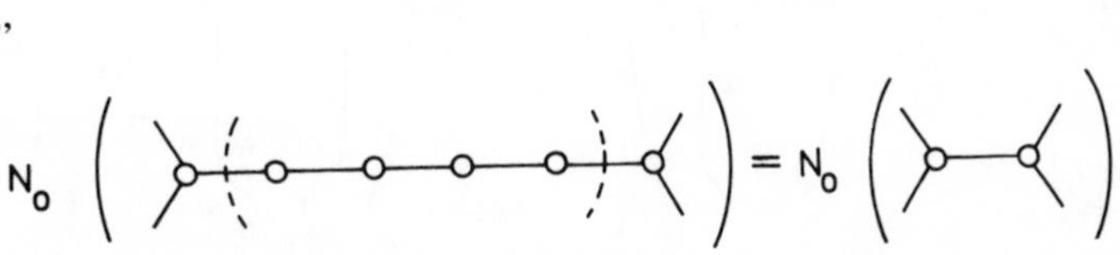

(36)

Example

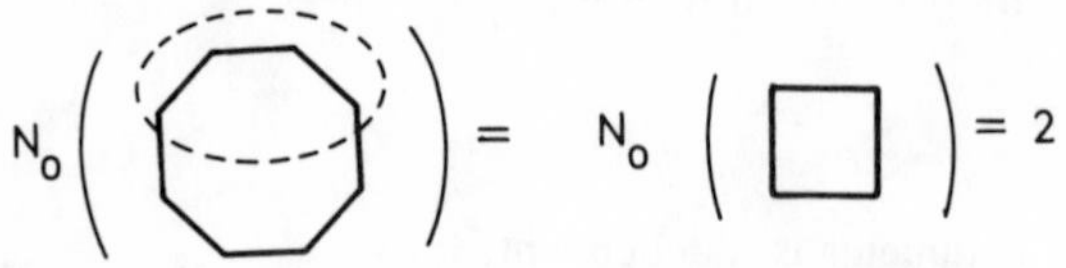

3. Two vertices and the adjacent edges plus the edge connecting the neighboring vertices of a peripheral four-membered ring can be removed without the changing the value of N_0,

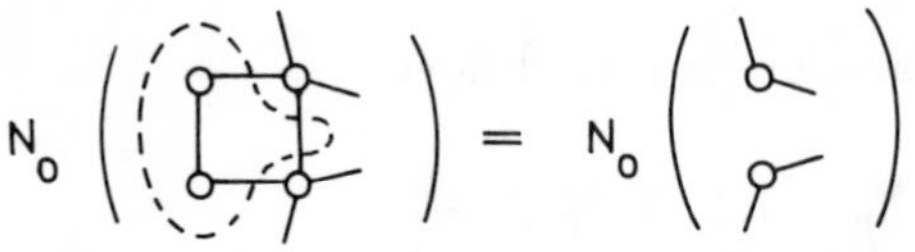

(37)

Example

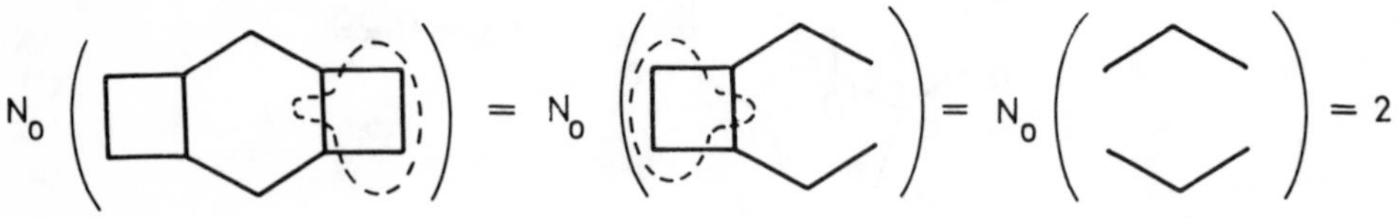

4. If the graph G has the following structure,

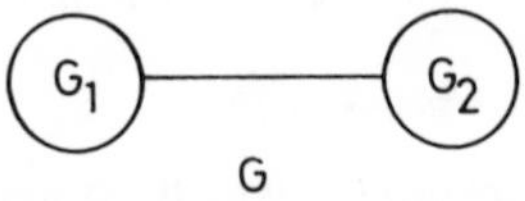

and if $N_0(G_1) = 0$, then $N_0(G) = N_0(G_2)$.

Example

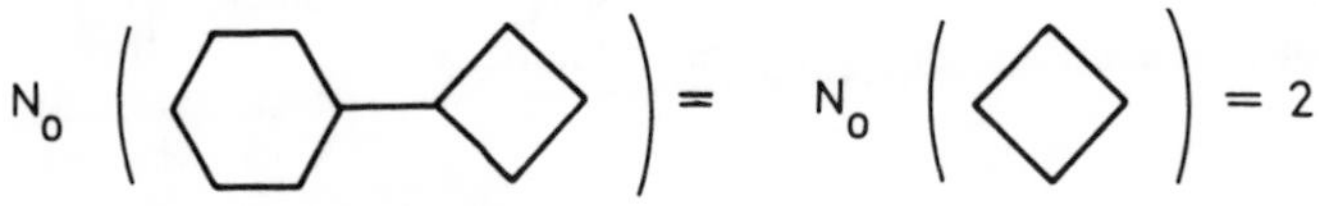

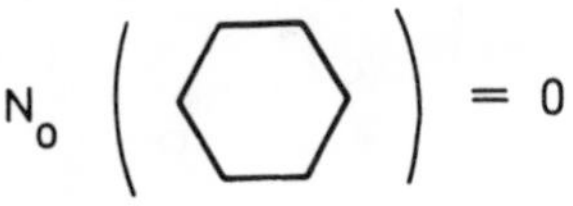

5. If a graph G consists of two parts G_i and G_j connected by a sequence of three edges, they can be replaced by a single edge without altering the value of N_0,

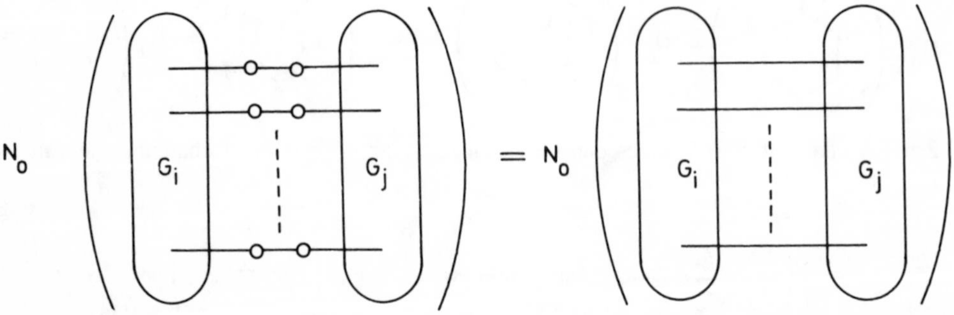

(38)

Example

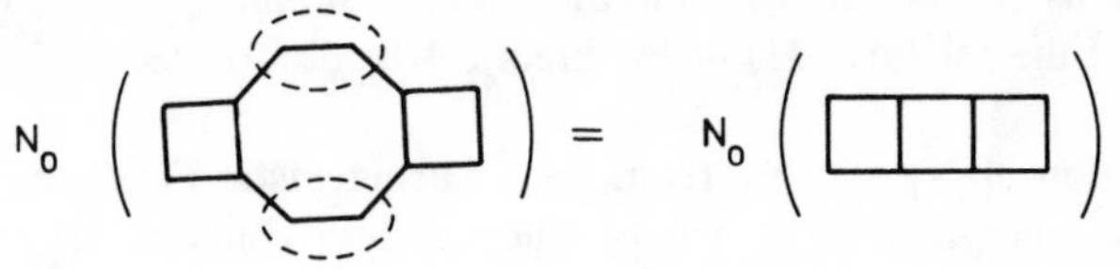

By applying the transformation (3), we arrive at the result that the above system has no NBMOs,

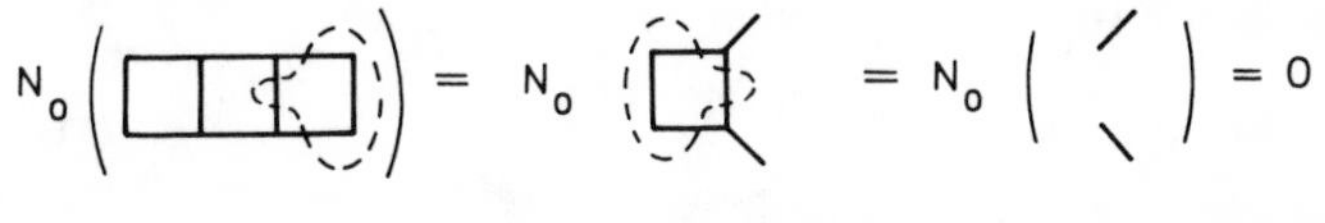

6. If G_i and G_j are bipartite fragments of G, connected in such a way that the edges connecting G_i and G_j join only the vertices of the same color in both subgraphs, then,

$$N_0(G) = N_0(G_i) + N_0(G_j) \quad (39)$$

Example

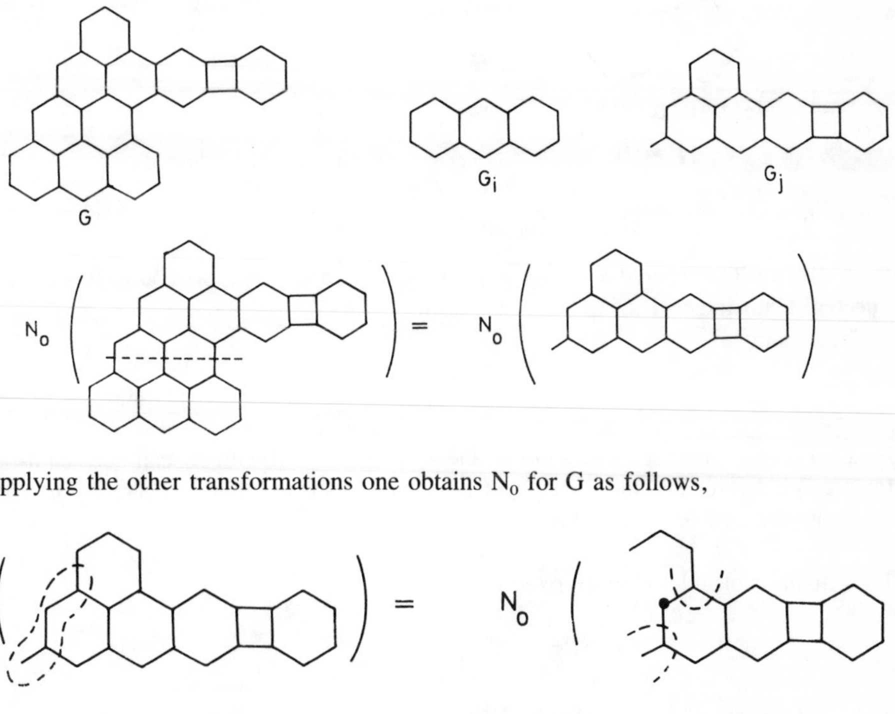

By applying the other transformations one obtains N_0 for G as follows,

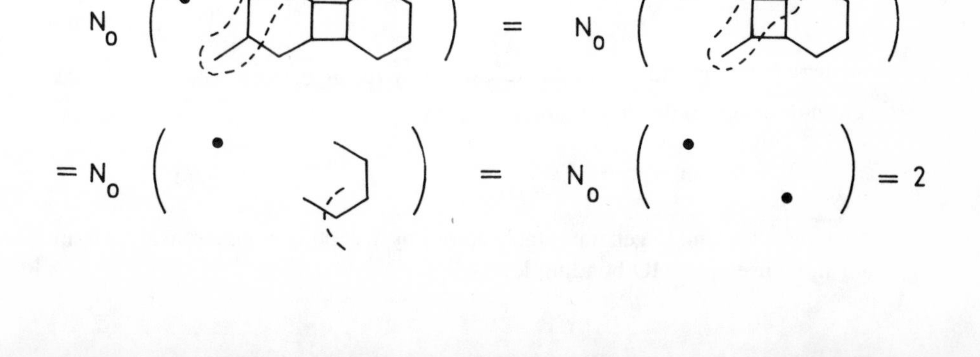

Graphical transformations, presented here, are very useful in the sense that by their use, quite complex graphs can be readily simplified. These simpler graphs can be then treated either by the use of the relation (31) or by direct calculation of det **A**.

C. The Enumeration of N_0 and N_+ from the Characteristic Polynomial

The characteristic polynomial of a conjugated system contains information about the magnitudes of numbers N_0 and N_+.[24,25] N_0 and N_+ are available directly from P(G; x) by using the following relations,

$$a_N = a_{N-1} = \dots = a_{N-N_0+1} = 0 \quad (40)$$

$$a_{N-N_0} \neq 0 \quad (41)$$

$$Ch(a_0, a_1, \dots, a_{N-1}, a_N) = N_+ \quad (42)$$

where **Ch** denotes the number of sign changes in the sequence of coefficients $a_j (j = 0, 1, 2, \dots, N)$. Equation (42) is known as *the Decartes theorem.*[26]

Example

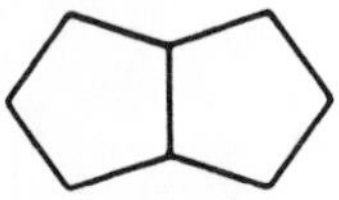

$$P(G; x) = x^8 - 9x^6 + 24x^4 - 4x^3 - 16x^2 + 8x$$

$a_8 = 0$, but $a_7 \neq 0$. Therefore

$8 = N - N_0 + 1$ or $N_0 = 1$

$Ch(1, -9, 24, -4, -16, 8) = N_+ = 4$

From Equation (28) follows, $N_- = 3$.

D. A Graph Theoretical Classification of Conjugated Hydrocarbons Based on their Spectral Characteristics

The quantity,

$$N_+ - N_- = \sigma \quad (43)$$

is called *a graph (molecule) signature* and when combined with N_0 can be used as a basis for a graph-theoretical classification of conjugated hydrocarbons.[27] There are *four* classes of conjugated hydrocarbons possible:

1. Stable molecules characterized by,

$$N_+ = N_- \text{ and } N_0 = 0 \quad (44)$$

2. Polyradical molecules characterized by,

$$N_+ = N_- \text{ and } N_0 > 0 \quad (45)$$

   Structures belonging to this class are extremely reactive.

3. Electron-deficient molecules characterized by,

$$N_+ > N_- \text{ and } N_0 \geqq 0 \quad (46)$$

   These molecules tend to generate stable anions by accepting π-electrons from a suitable donor in their empty MO bonding levels.[28]

4. Electron-excessive molecules characterized by,

$$N_+ < N_- \text{ and } N_0 \geqq 0 \quad (47)$$

These molecules show a tendency to generate stable cations by releasing electrons from the antibonding MO levels.[28]

Pentalene, studied earlier in this chapter (Section III.C), belongs to class 3. It is a reactive electron-deficient molecule[28] whose dianion is relatively stable.[29]

Polycyclic hydrocarbons consisting of $(4m + 2)$- and/or $(4m)$-rings belong either to class 1 or class 2. The presence of a $(4m - 1)$-membered ring in a graph is a necessary topological condition for a corresponding molecule to belong to class 3, while the $(4m + 1)$-ring is required for class 4.

IV. CHARGE DENSITIES AND BOND ORDERS IN CONJUGATED SYSTEMS

It was Coulson[30] who first defined charge densities and bond orders in Hückel theory. An element in the Coulson charge density-bond order matrix $\mathbf{P}^C$ is given by,

$$(\mathbf{P}^C)_{rs} = \sum_{i=1}^{N} g_i \, c_{ir} \, c_{is} \quad (48)$$

where c_{ir} and c_{is} are linear expansion coefficients defining the contributions of the r-th and s-th atomic orbitals, respectively, in the i-th MO.

Alternative definitions, within the MO theory, of charge densities and bond orders were offered by Mulliken[31] and Ruedenberg.[11] Elements of the Mulliken, $\mathbf{P}^M$, and the Ruedenberg, $\mathbf{P}^R$, charge density-bond order matrices are defined as,

$$(\mathbf{P}^M)_{rs} = (1 + S_{rs}) \sum_{i=1}^{N} g_i \frac{c_{ir} \, c_{is}}{1 + x_i \, S_{rs}} \quad (49)$$

$$(\mathbf{P}^R)_{rs} = \sum_{i=1}^{N} g_i \frac{c_{ir} \, c_{is}}{x_i} \quad (50)$$

The formulae by Coulson, Mulliken, and Ruedenberg can be collected together in the general charge density-bond order matrix,

$$(\mathbf{P})_{rs} = \sum_{i=1}^{N} g_i \, c_{ir} \, c_{is} \, f(x_i) \quad (51)$$

where $f(x_i)$ is a weighting factor. An analysis by Ham and Ruedenberg[32] indicated that there is a connection between the Coulson and Mulliken charge density-bond order matrix,

$$\left[(\mathbf{P}^M)_{rs} - 1 \right] \approx 1.2 \left[(\mathbf{P}^C)_{rs} - 1 \right] \quad (52)$$

In addition, they have shown that the charge density-bond order matrix $\mathbf{P}$ may be expressed in terms of the adjacency matrix of the molecule,

$$\mathbf{P} = \mathbf{P}^C \, f(\mathbf{A}) \quad (53)$$

where

$$\mathbf{P}^C = \mathbf{I} + (\mathbf{A}^2)^{1/2} \, \mathbf{A}^{-1} \quad (54)$$

Since the bond orders may be correlated with bond lengths,[30,33] the important conclusion of these considerations is that molecular topology in a subtle way influences the geometry of a conjugated system.

Pauling and co-workers[34] have defined a bond order within the resonance theory, a simple version of the valence bond theory, as,

$$(p^P)_{rs} = \frac{K[G-(r-s)]}{K(G)} \tag{55}$$

where K(G) and K[G - (r - s)] are the numbers of 1-factors (Kekulé structures) for a graph G (molecule) and subgraph G - (r - s) (molecular fragment) obtained by deletion of the vertices *r* and *s* (atoms) and the adjacent edges (bonds) from G (molecule). The above formulation of $\mathbf{P}^P$ is given in the graph-theoretical formalism.[35-37] The definition of the Pauling bond order precedes the definitions in the framework of MO theory.

Example

$K(G) = 7$ $\qquad$ $K[G-(r-s)] = 4$

$$(p^P)_{rs} = \frac{4}{7} = 0.57$$

If the subgraph G - (r - s) has no Kekulé structure, then the corresponding bond $\ell_{r\text{-}s}$ has a bond order of zero.

An interesting result is obtained when $\mathbf{P}^R$ and $\mathbf{P}^P$ are compared. It is found that for alternant hydrocarbons without 4m-membered rings and NBMOs, bond orders defined by Ruedenberg are *identical* with the Pauling bond orders,[38]

$$p^R = p^P \tag{56}$$

Example

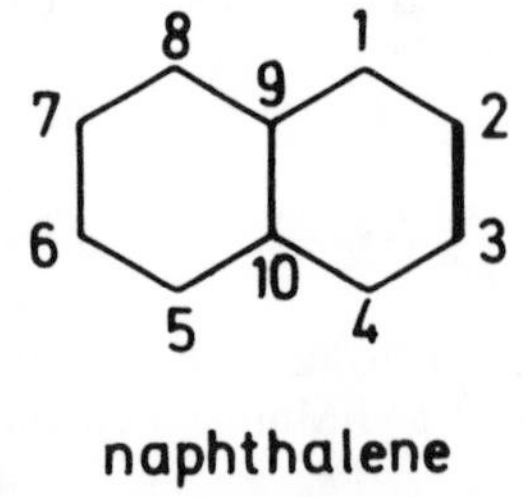

$$(p^R)_{23} = 2\left(\frac{c_{12}\,c_{13}}{x_1} + \frac{c_{22}\,c_{23}}{x_2} + \frac{c_{32}\,c_{33}}{x_3} + \frac{c_{42}\,c_{43}}{x_4} + \frac{c_{52}\,c_{53}}{x_5}\right) +$$

$$0\left(\frac{c_{62}\,c_{63}}{x_6} + \frac{c_{72}\,c_{73}}{x_7} + \frac{c_{82}\,c_{83}}{x_8} + \frac{c_{92}\,c_{93}}{x_9} + \frac{c_{10,2}\,c_{10,3}}{x_{10}}\right) =$$

$$2\left(\frac{0.230\cdot 0.230}{2.303} + \frac{0.425\cdot 0.425}{1.618} + \frac{0.174\,(-0.174)}{1.303} +\right.$$

$$\left.\frac{(-0.408)\,(-0.408)}{1.000} + \frac{0.263\,(-0.263)}{0.618}\right) = 0.33$$

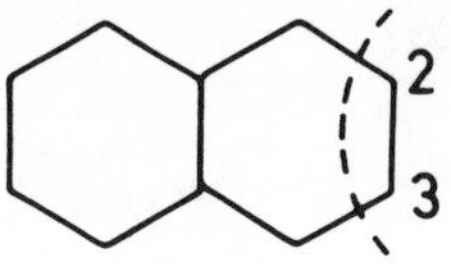

G

$K(G) = 3$

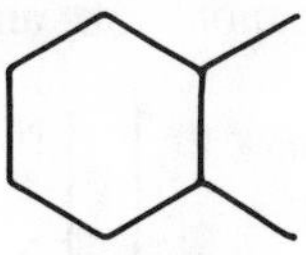

G-(2-3)

$K[G-(2-3)] = 1 \qquad (\mathbf{P}^{P})_{23} = 1/3 = 0.33$

These results show that we can use the tables of Hückel values[10] to evaluate Pauling bond orders. More important is the conclusion that both HMO and Pauling concepts of bond order are, in essence, topological quantities.

V. THE TWO-COLOR PROBLEM IN HÜCKEL THEORY

Conjugated molecules that can be represented by the bipartite graphs (the two-colored structures) are called *alternant hydrocarbons* (AHs).[39] *Nonalternant hydrocarbons* are depicted by nonbipartite graphs.

A fact that the adjacency matrix belonging to a bipartite graph appears in the block form (see Equation 15, Chapter 4) may be used in discussing the pairing theorem of Coulson and Rushbrooke.[40] According to the pairing theorem, if x_i is an element of a Hückel spectrum (with an associated eigenvector $\psi_i = \sum_r c^*_{ir}\, \phi_r + \sum_s c^0_{is}\, \phi_s$), then $-x_i$ (with an associated eigenvector $\psi_i = \sum_r c^*_{ir}\, \phi_r - \sum_s c^0_{is}\, \phi_s$) is also an eigenvalue of $\mathbf{A}(G)$. In other words, the pairing theorem states that Hückel energy levels of an alternant hydrocarbon should be symmetrically distributed about $x = 0$. This result was also observed by Hückel[3] 8 years before the work by Coulson and Rushbrooke. Since the Hückel eigenvalues for alternants appear in pairs: $x_i + x_{N+1-i} = 0$, in order to construct the whole set of the HMO levels and MOs it is sufficient to obtain, in some way, only positive (or negative) eigenvalues and their corresponding eigenvectors.[41]

There are a number of proofs and demonstration of the pairing theorem available in the literature.[5-8,10,18,40-46] Here we will briefly discuss the matrix formulation of the spectral symmetry.

Let the eigenvalue equation,

$$\mathbf{C}_i\, \mathbf{A} = x_i\, \mathbf{C}_i \qquad (57)$$

be satisfied for the eigenvector,

$$\mathbf{C}_i = \left[c^*_{i1}, c^*_{i2}, \ldots, c^*_{is}, c^0_{i,s+1}, c^0_{i,s+2}, \ldots, c^0_{i,s+u}\right] \qquad (58)$$

There should be another eigenvalue equation, producing the other member of the pair: x_i, $-x_i$,

$$\mathbf{C}^{pair}_i\, \mathbf{A} = -x_i\, \mathbf{C}^{pair}_i \qquad (59)$$

with the eigenvector,

$$\mathbf{C}^{pair}_i = \left[c^*_{i1}, c^*_{i2}, \ldots, c^*_{is}, -c^0_{i,s+1}, -c^0_{i,s+2}, \ldots, -c^0_{i,s+u}\right] \qquad (60)$$

In the matrix notation Equations (58) and (60) are given by,

$$\mathbf{C} = \begin{bmatrix} C_{\text{bonding}} \\ \\ C_{\text{antibonding}} \end{bmatrix} = \begin{bmatrix} \mathbf{N} & \mathbf{M} \\ \\ \mathbf{N} & -\mathbf{M} \end{bmatrix} \tag{61}$$

where

$$\mathbf{N} = \left[c^*_{i1}, c^*_{i2}, \ldots, c^*_{is}\right] \tag{62}$$

$$\mathbf{M} = \left[c^0_{i,s+1}, c^0_{i,s+2}, \ldots, c^0_{i,s+u}\right] \tag{63}$$

Since the adjacency matrix of an alternant structure can, by an appropriate labeling of atoms (vertices), be given in the block form, the eigenvalue matrix **X** is expressed as follows,

$$\mathbf{X} = \mathbf{C}\,\mathbf{A}\,\mathbf{C}^{\dagger} = \begin{bmatrix} \mathbf{Y} & 0 \\ \\ 0 & -\mathbf{Y} \end{bmatrix} \tag{64}$$

where

$$\mathbf{Y}/2 = \mathbf{N}\,\mathbf{B}\,\mathbf{M}^{\dagger} = \mathbf{M}\,\mathbf{B}^{\dagger}\,\mathbf{N}^{\dagger} \tag{65}$$

where **B** is a submatrix of **A,** while **Y** is a diagonal matrix. Therefore, the symmetry of Hückel levels is the very consequence of the particular topology of alternant systems.

Another way of introducing the pairing theorem is by using the Sachs formula. Because the bipartite graphs, by a theorem, do not contain odd-membered cycles, it follows that for alternant structures $S_n = 0$ and $a_n(G) = 0$ for $n = 2j + 1$. Therefore, the characteristic polynomial of alternant hydrocarbons is necessarily of the form,

$$P(G; x) = \sum_{j=0}^{[N/2]} (-1)^j\, a_{2j}(G)\, x^{N-2j} \tag{66}$$

where the coefficients $a_{2j}(G)$ are the nonnegative quantities for all j. The structure of the polynomial (66) is such that the corresponding spectrum *must* be symmetrically arranged with respect to $x = 0$, because if x is a root of a polynomial (66), then $-x$ must be also a root. This is possibly the simplest demonstration of the pairing theorem.

The pairing theorem may be extended to include other bicolorable structures encountered in chemistry. However, in these structures the starred and unstarred positions are populated with different types of atoms. Here we give a theorem which is valid for alternant hetero-conjugated structures, like *s*-triazine,

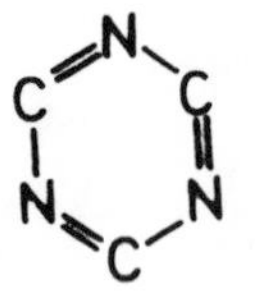

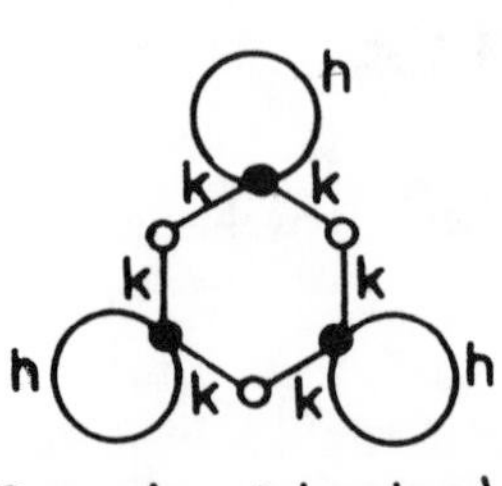

s-triazine $\qquad$ G_{VEW}(s-triazine)

The extension of the pairing theorem to include bipartite graphs of the type G_{VEW} is stated as follows:[47] if G_{VEW} is a bipartite graph with the same number of vertices in each set, and exactly those in the first are weighted, then,

$$x_i + x_{N+1-i} = h \quad \text{for} \quad 1 \leqslant i \leqslant N \tag{67}$$

To illustrate the use of the above theorem let us consider *s*-triazine system. Its characteristic polynomial is given by,

$$P(G_{VEW}; x) = x^6 - 3hx^5 + (3h^2 - 6k^2)x^4 - (12hk^2 - h^3)x^3 + (9k^4 - 6h^2)x^2 - 9hk^4x - 4k^6 \tag{68}$$

Taking, for example, $h = 2.00$ and $k = 1.00$, the following spectrum is obtained,

$$\{3.23607, 2.41421, 2.41421, -0.41421, -0.41421, -1.23607\} \tag{69}$$

The use of Equation (67) shows that the above spectrum is symmetric about $h/2 = 1.00$. For example,

$$\tfrac{1}{2}(x_2 + x_5) = \tfrac{1}{2}(2.41421 - 0.41421) = 1.00 \tag{70}$$

A. Properties of Alternant Hydrocarbons

There are some unique properties of AHs which can be easily explained in terms of the pairing theorem.

If NBMOs appear in the Hückel spectrum of *even* AHs, their number will be always *even,* because all elements of the spectrum must be paired. However, in the spectrum of odd AHs there is always at least *one* NBMO present, because $N - 1$ levels will be symmetrically arranged around $x = 0$, and this leaves a single level which must be a NBMO.

Even alternant hydrocarbons

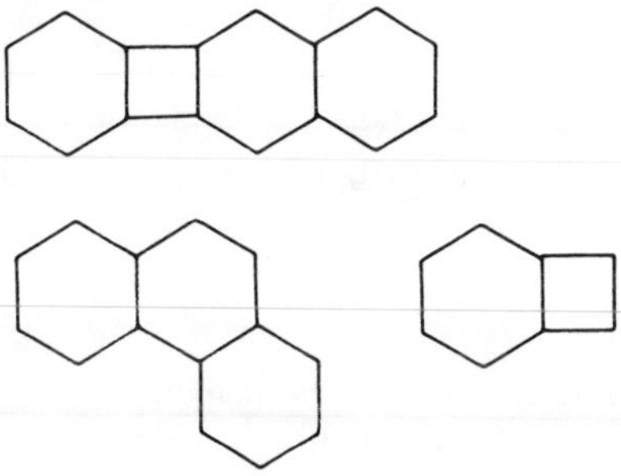

Odd alternant hydrocarbons

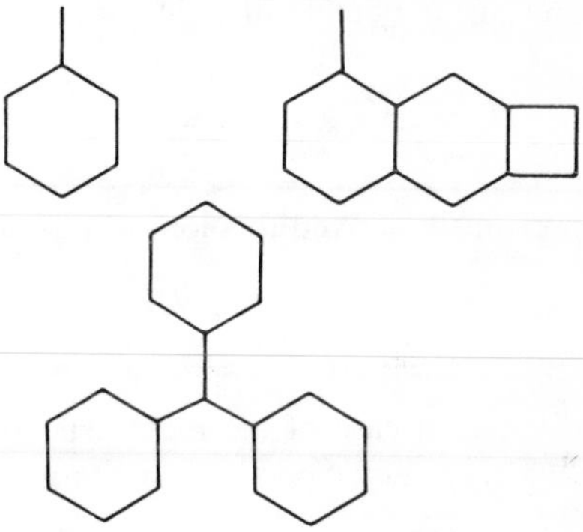

Another consequence of the pairing theorem is the fact that the charge density distribution in alternant hydrocarbons is uniform and that each atom carries the π-electron density value equal unity.[10] This can be proved in the following way. The Coulson charge density-bond order matrix is given by,

$$\mathbf{P}^C = 2\,\mathbf{C}^{\dagger}_{\text{bonding}}\,\mathbf{C}_{\text{bonding}} \tag{71}$$

where

$$\mathbf{C}_{\text{bonding}} = [\mathbf{N}, \mathbf{M}] \tag{72}$$

Applying the relations,

$$\mathbf{C}^{\dagger}\mathbf{C} = \mathbf{I} \tag{73}$$

$$2\mathbf{N}^{\dagger}\mathbf{N} = 2\,\mathbf{M}^{\dagger}\mathbf{M} \tag{74}$$

matrix (71) may be given in the form,

$$\mathbf{P}^C = \begin{bmatrix} \mathbf{I} & 2\mathbf{N}^\dagger\mathbf{M} \\ 2\mathbf{M}^\dagger\mathbf{N} & \mathbf{I} \end{bmatrix} \tag{75}$$

From this matrix is seen that the π-electron charge density on any arbitrary atom r of an AH is identically equal unity,

$$q_r = (\mathbf{P}^C)_{rr} = 1 \tag{76}$$

The above results leads to the prediction of a zero π-component of dipole moment in AHs. Available experimental data verify this prediction; AHs have indeed either zero or negligibly small total dipole moments.[48]

The fact that the charge density distribution is uniform is related to the self-consistency of the topological orbitals of alternant systems. It has been shown[49] that this self-consistency of topological MOs depends on the relations,

$$(\mathbf{P}^C)_{11} = (\mathbf{P}^C)_{22} = \ldots = (\mathbf{P}^C)_{rr} = \ldots = (\mathbf{P}^C)_{NN} \tag{77}$$

An important property of AHs is that the bond orders between atoms of the same color are zero,

$$(\mathbf{P}^C)_{**} = (\mathbf{P}^C)_{oo} = 0 \tag{78}$$

The nonvanishing bond orders in AHs are possible only between differently colored atoms,

$$(\mathbf{P}^C)_{*o} = 2\mathbf{N}^\dagger\mathbf{M} \tag{79}$$

A previously derived topological formula for bond order by Hall,[42]

$$(\mathbf{P}^C)_{*o} = (\mathbf{B}\mathbf{B}^\dagger)^{-1/2}\,\mathbf{B} \tag{80}$$

is just a special case of the more general formula (54). This can be demonstrated in the following way. Equation (54) may be given in a more convenient form,

$$\mathbf{P}^C = \begin{bmatrix} \mathbf{I} & \mathbf{B}\,(\mathbf{B}^\dagger\mathbf{B})^{-1/2} \\ \mathbf{B}^\dagger\,(\mathbf{B}\,\mathbf{B}^\dagger)^{-1/2} & \mathbf{I} \end{bmatrix} \tag{81}$$

where use is made of the matrices,

$$\mathbf{A}^{-1} = \begin{bmatrix} \mathbf{0} & (\mathbf{B}^\dagger)^{-1} \\ \mathbf{B}^{-1} & \mathbf{0} \end{bmatrix} \tag{82}$$

and

$$(\mathbf{A}^2)^{1/2} = \begin{bmatrix} (\mathbf{B}\,\mathbf{B}^\dagger)^{1/2} & \mathbf{0} \\ \mathbf{0} & (\mathbf{B}^\dagger\,\mathbf{B})^{1/2} \end{bmatrix} \tag{83}$$

Equation (81) immediately enables one to obtain results (76), (78), and (79). In the case of NAHs, the matrix $[(\mathbf{A}^2)^{1/2}\,\mathbf{A}^{-1}]$ has nonvanishing diagonal elements; the π-electron charge density is thus nonuniform, and dipole moments are nonzero.

VI. EIGENVALUES OF LINEAR POLYENES

A linear polyene, [N]-polyene, of the general formula $CH_2(CH)_{N-2}CH_2$, may be represented by chain of N vertices,

L_N

The corresponding adjacency matrix is of the form,

$$A(L_N) = \begin{array}{c} \\ 1 \\ 2 \\ 3 \\ \cdot \\ \cdot \\ \cdot \\ \cdot \\ N-2 \\ N-1 \\ N \end{array} \begin{array}{c} \begin{array}{cccccccc} 1 & 2 & 3 & \cdots & & N-2 & N-1 & N \end{array} \\ \left[\begin{array}{cccccccc} 0 & 1 & 0 & & & & & \\ 1 & 0 & 1 & & & & \mathbf{0} & \\ & 1 & 0 & 1 & & & & \\ & & 1 & 0 & 1 & & & \\ & & & & \ddots & & & \\ & \mathbf{0} & & & & & & \\ & & & & 1 & 0 & 1 & \\ & & & & & 1 & 0 & 1 \\ & & & & & & 1 & 0 \end{array}\right] \end{array} \tag{84}$$

This matrix reflects clearly the topology of the chain, nonzero entries appearing only alongside the principal diagonal. Since it is known that $|x| \leq 2$ (see Chapter 5, Section I), one can formally substitute $x = 2\cos\theta$ in $\det|x\mathbf{I} - \mathbf{A}(L_N)|$. The expansion of the determinant gives the following characteristic polynomial,[50]

$$P(L_N; x) = 2\cos\theta \, P(L_{N-1}; x) - P(L_{N-2}; x) \tag{85}$$

It can be shown by induction that,

$$P(L_N; x) = \frac{\sin(N+1)\theta}{\sin\theta} \tag{86}$$

The above polynomial vanishes when and only when,

$$\theta = \frac{j\pi}{N+1} \qquad (j = 1,2,\ldots,N) \tag{87}$$

that is, when,

$$x_j = 2\cos\frac{j\pi}{N+1}; j = 1,2,\ldots,N \tag{88}$$

These are in fact the N distinct zeroes of the polynomial $P(L_N; x)$. Thus, the eigenvalues of linear polyenes are obtainable from the relation,

$$E_j = x_j = 2\cos\frac{j\pi}{N+1} \qquad (j = 1,2,\ldots,N) \tag{89}$$

There are N distinct roots (Hückel eigenvalues) of the equation $P(L_N; x) = 0$. Since $P(L_N; x)$ is a polynomial in x of degree N and since the coefficient of x^N in this polynomial is +1 it follows,

$$P(L_N; x) = \prod_{j=1}^{N} \left(x - 2\cos\frac{j\pi}{N+1}\right) \tag{90}$$

Example

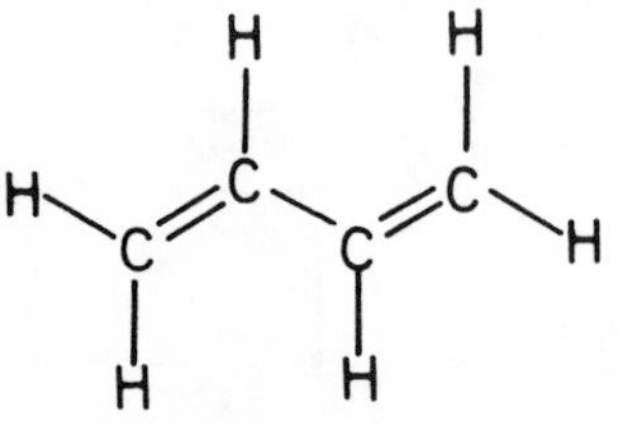

Butadiene

L_4

N = 4

$$E_1 = 2\cos\frac{\pi}{5} = 1.6180$$

$$E_2 = 2\cos\frac{2\pi}{5} = 0.6180$$

$$E_3 = 2\cos\frac{3\pi}{5} = -0.6180$$

$$E_4 = 2\cos\frac{4\pi}{5} = -1.6180$$

$$P(L_4; x) = (x - 1.6180)(x - 0.6180)(x + 0.6180)(x + 1.6180) =$$

$$x^4 - 3x^2 + 1$$

A very useful mnemonic device for determining the energy level diagrams of linear polyenes has been found by Frost and Musulin.[51] They have noticed that one can derive the Hückel energies for regular cyclic polyenes by inscribing the appropriate regular hexagon in a cycle of radius 2 with one vertex down. The cycle is centered at $E_i = 0$. For every intersection of the cycle and polygon, there is an MO energy level corresponding to the vertical displacement. Since every polygon has one vertex down and intersecting at the bottom of the cycle, there always will be an Hückel level of 2, i.e., the value corresponding to the cycle radius. The horizontal projection of all other points of contact onto a vertical line yield the exact eigenvalues for the cyclic molecule. In order to obtain the energy levels of a given [N]-polyene, a fictitious [N + 2]-polyene is added on a [2N + 2]-polygon in such a way that the first vertex of the polyene graph touches the second vertex of the polygon. The use of the Frost-Musulin device is illustrated in Figure 2.

We note that all eigenvalues appear in pairs. This is yet another demonstration of the pairing theorem. An additional interesting observation is that the Hückel spectrum of [10]-annulene contains twice the entire spectrum of [4]-polyene. Since we know from above the spectrum of [4]-polyene, we can immediately put down the spectrum of [10]-annulene,

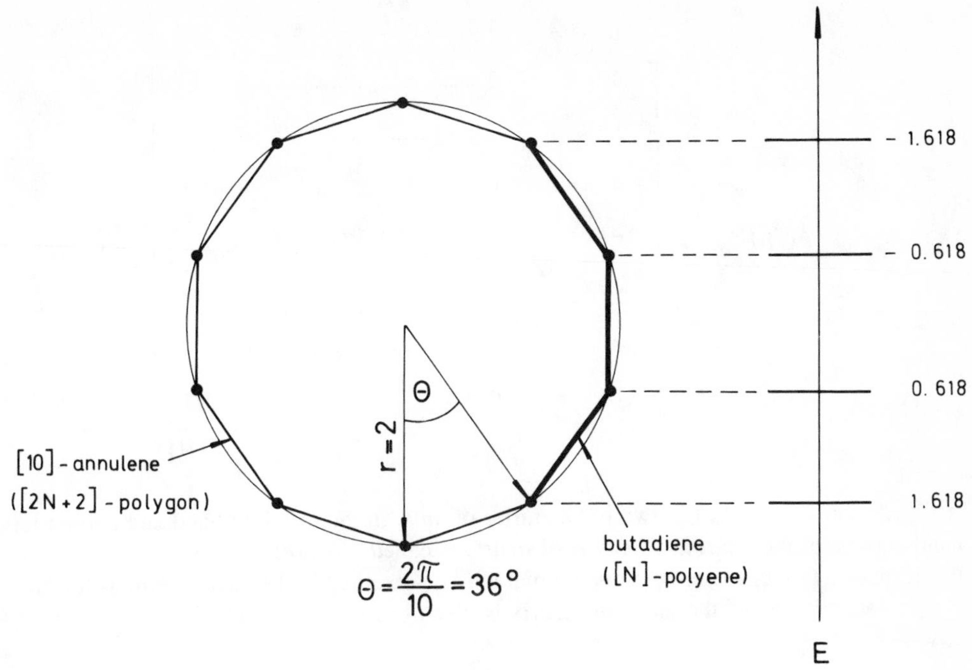

FIGURE 2. The Frost-Musulin projection diagram for the eigenvalues of butadiene.

$$\{2,\ 1.618,\ 1.618,\ 0.618,\ 0.618,\ -\ 0.618,\ -0.618,\ -\ 1.618,\ -2\}$$

Such a relationship between the spectra of a graph and its subgraph is called *subspectrality* and it will be discussed in Chapter 8.

VII. EIGENVALUES OF ANNULENES

A cyclic polyene, [N]-annulene,[52] of the general formula $(CH)_N$ may be represented by a cycle of N vertices,

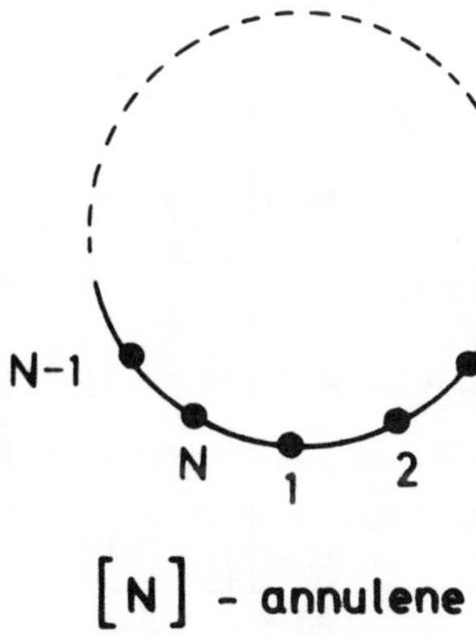

The corresponding adjacency matrix is very similar to that belonging to [N]-polyene,[53]

$$
\mathbf{A}(C_N) = \begin{array}{c} \\ 1 \\ 2 \\ 3 \\ \cdot \\ \cdot \\ N-1 \\ N \end{array}
\begin{array}{c} \begin{array}{cccccc} 1 & 2 & 3 & \cdots\cdots & N-1 & N \end{array} \\
\left[\begin{array}{cccccc}
0 & 1 & 0 & \cdots\cdots & 0 & \textcircled{1} \\
1 & 0 & 1 & \cdots\cdots & 0 & 0 \\
0 & 1 & 0 & \cdots\cdots & 0 & 0 \\
\cdot & \cdot & \cdot & \cdot\ \cdot\ \cdot & \cdot & \cdot \\
\cdot & \cdot & \cdot & \cdot\ \cdot\ \cdot & \cdot & \cdot \\
0 & 0 & 0 & \cdots\cdots & 0 & 1 \\
\textcircled{1} & 0 & 0 & \cdots\cdots & 1 & 0
\end{array}\right] \end{array} \tag{91}
$$

$\mathbf{A}(C_N)$ differs from $\mathbf{A}(L_N)$ in two extra entries of *unity* in the top right hand and bottom left hand corners of the matrix. This type of matrix is called *circulant matrix*.[54,55] The zeros of the corresponding characteristic polynomial may be obtained in the closed form as follows.

The general form of the circulant matrix is given by,

$$
\left[\begin{array}{ccccccc}
a_1 & a_2 & a_3 & \cdots\cdots & & & a_N \\
a_N & a_1 & a_2 & \cdots\cdots & & & a_{N-1} \\
a_{N-1} & a_N & a_1 & \cdots\cdots & & & a_{N-2} \\
\cdot & \cdot & \cdot & \cdot & \cdot & \cdot & \cdot \\
\cdot & \cdot & \cdot & \cdot & \cdot & \cdot & \cdot \\
a_2 & a_3 & a_4 & & & & a_1
\end{array}\right] \tag{92}
$$

The circulant matrix is completely defined when the elements in its first row are specified. The diagonalization of the matrix (92) produces a set of eigenvectors and eigenvalues,

$$
\left[\begin{array}{ccccc}
a_1 & a_2 & a_3 & \cdots\cdots & a_N \\
a_N & a_1 & a_2 & \cdots\cdots & a_{N-1} \\
a_{N-1} & a_N & a_1 & \cdots\cdots & a_{N-2} \\
\cdot & \cdot & \cdot & \cdot & \cdot \\
\cdot & \cdot & \cdot & \cdot & \cdot \\
a_2 & a_3 & a_4 & \cdots\cdots & a_1
\end{array}\right]
\left[\begin{array}{c} 1 \\ \rho_j \\ \rho_j^2 \\ \cdot \\ \cdot \\ \cdot \\ \rho_j^{N-1} \end{array}\right]
= x_j
\left[\begin{array}{c} 1 \\ \rho_j \\ \rho_j^2 \\ \cdot \\ \cdot \\ \cdot \\ \rho_j^{N-1} \end{array}\right] \tag{93}
$$

Thus, the vector

$$\mathbf{V}_k = \begin{bmatrix} 1 \\ \rho_j \\ \rho_j^2 \\ \cdot \\ \cdot \\ \cdot \\ \rho_j^{N-1} \end{bmatrix} \tag{94}$$

is an eigenvector of the circulant matrix (92) with the eigenvalue x_j given by,

$$x_j = a_1 \rho_j^0 + a_2 \rho_j^1 + a_3 \rho_j^2 + \ldots\ldots + a_N \rho_j^{N-1} \tag{95}$$

where ρ_j is an N-th root of the scalar equation,

$$\rho_j^N = 1;\ j = 0, 1, 2, \ldots\ldots, N - 1 \tag{96}$$

The solutions of the above equation are as follows,

$$\rho_j = \cos\left(\frac{2j\pi}{N}\right) + i \sin\left(\frac{2j\pi}{N}\right);\ j = 0, 1, 2, \ldots\ldots, N - 1 \tag{97}$$

Going back to the circulant matrix of the [N]-annulene (91) we note that all elements in first row vanish, but a_2 and a_N, both being equal to unity. Thus, using this result, Equation (95) reduces to,

$$x_j = \rho_j + \rho_j^{N-1} \tag{98}$$

Since matrix (91) is a Hermitian matrix,

$$\rho_j^{N-1} = \rho_j^{\star} \tag{99}$$

it follows that,

$$x_j = \rho_j + \rho_j^{\star} \tag{100}$$

or by substituting (97) for ρ_j,

$$x_j = 2 \cos\left(\frac{2j\pi}{N}\right) \tag{101}$$

Because of (17),

$$E_j = 2 \cos\left(\frac{2j\pi}{N}\right);\ j = 0, 1, 2, \ldots\ldots, N - 1 \tag{102}$$

This is equation for calculating the eigenvalues of [N]-annulenes.

Example

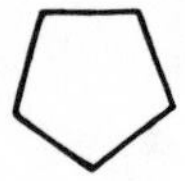

[5] - annulene

$$E_0 = 2\cos 0^0 = 2$$

$$E_1 = 2\cos\frac{2\pi}{5} = 0.618 = E_4 = 2\cos\frac{8\pi}{5}$$

$$E_2 = 2\cos\frac{4\pi}{5} = -1.618 = E_3 = 2\cos\frac{6\pi}{5}$$

A mnemonic device of Frost and Musulin[51] may be used as described in Section VI for constructing the energy level diagrams of [N]-annulenes. This is demonstrated for [5]- and [10]-annulene in Figure 3.

We note the following. When N = even, there will be two nondegenerate levels, +2 for $j = 0$ and -2 for $j = N/2$. All the others will occur in degenerate pairs. This is another demonstration of the applicability of the pairing theorem. When N = odd, there will be a unique energy level of 2 units for j = 0. All the remaining energy levels will appear in degenerate pairs.

VIII. EIGENVALUES OF MÖBIUS ANNULENES

A Mobius [N]-annulene is a cyclic structure with, at least, one phase dislocation as described in Chapter 3, Section VI. The Möbius [N]-annulene may be represented graph-theoretically by a Möbius cycle of N vertices,[56]

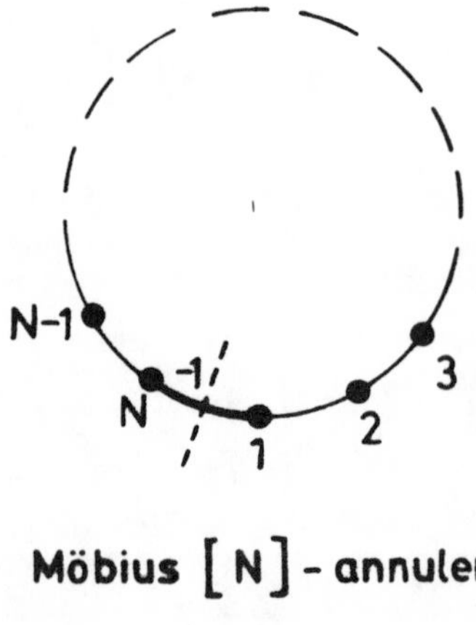

Möbius [N] - annulene

The corresponding adjacency matrix is very similar to that of [N]-cycle,

$$
\mathbf{A}\left(C_N^{M\ddot{o}}\right) = \begin{array}{c} \\ 1 \\ 2 \\ 3 \\ \cdot \\ \cdot \\ \cdot \\ N-1 \\ N \end{array} \begin{array}{c} \begin{array}{cccccccc} 1 & 2 & 3 & & & & N-1 & N \end{array} \\ \left[\begin{array}{cccccccc} 0 & 1 & 0 & & & & 0 & \circled{-1} \\ 1 & 0 & 1 & & & & 0 & 0 \\ 0 & 1 & 0 & & & & 0 & 0 \\ \cdot & \cdot & \cdot & \cdot & \cdot & & \cdot & \cdot \\ \cdot & \cdot & \cdot & \cdot & \cdot & & \cdot & \\ \cdot & \cdot & \cdot & \cdot & \cdot & & \cdot & \\ 0 & 0 & 0 & & & & 0 & 1 \\ \circled{-1} & 0 & 0 & & & & 1 & 0 \end{array}\right] \end{array} \tag{103}
$$

$\mathbf{A}(C_N^{M\ddot{o}})$ differs from $\mathbf{A}(C_N)$ in the sign of the two entries in the top right hand and bottom left hand corners of the matrix.

The eigenvalues of the Möbius [N]-annulenes may be obtained in the closed form using similar reasoning as in the case of [N]-annulenes because matrix (103) is a special case of the circulant matrix. The expression for calculating the eigenvalues of Möbius cycles is given by,[57]

$$
E_j = 2\cos\frac{(2j+1)\pi}{N}\ ;\ j = 0, 1, \ldots, N-1 \tag{104}
$$

Example

Möbius [5]-annulene

$$E_1 = 2\cos\frac{\pi}{5} = 1.618 = E_5 = 2\cos\frac{9\pi}{5}$$

$$E_2 = 2\cos\frac{3\pi}{5} = -0.618 = E_4 = 2\cos\frac{7\pi}{5}$$

$$E_3 = 2\cos\frac{5\pi}{5} = -2.00$$

A Möbius mnemonic device may be introduced following the ideas of Frost and Musulin.[51] This was actually done by Zimmerman.[58] In order to obtain the Möbius-Zimmerman projection diagrams a vertex at the bottom is replaced by the side of the polygon. Thus, the first two eigenvalues of the Möbius annulene will always be a degenerate pair. In the case of the Möbius annulenes with N = even, all energy levels will appear in the degenerate pairs. An additional feature appears in the Möbius annulenes with N/2 = odd, because they will always have a degenerate pair with zero eigenvalue. In the case of Möbius annulenes with N = odd, N − 1 levels will appear in degenerate pairs. Last energy level will be a single one with the value −2.00.

The application of the Möbius-Zimmerman device to Möbius [5]- and [10]-annulenes is demonstrated in Figure 4.

The inspection of Figure 4 reveals that the pairing theorem holds in the Möbius systems with N = even. This is one of the topological effects that operates in both Möbius and Hückel systems with identical result.

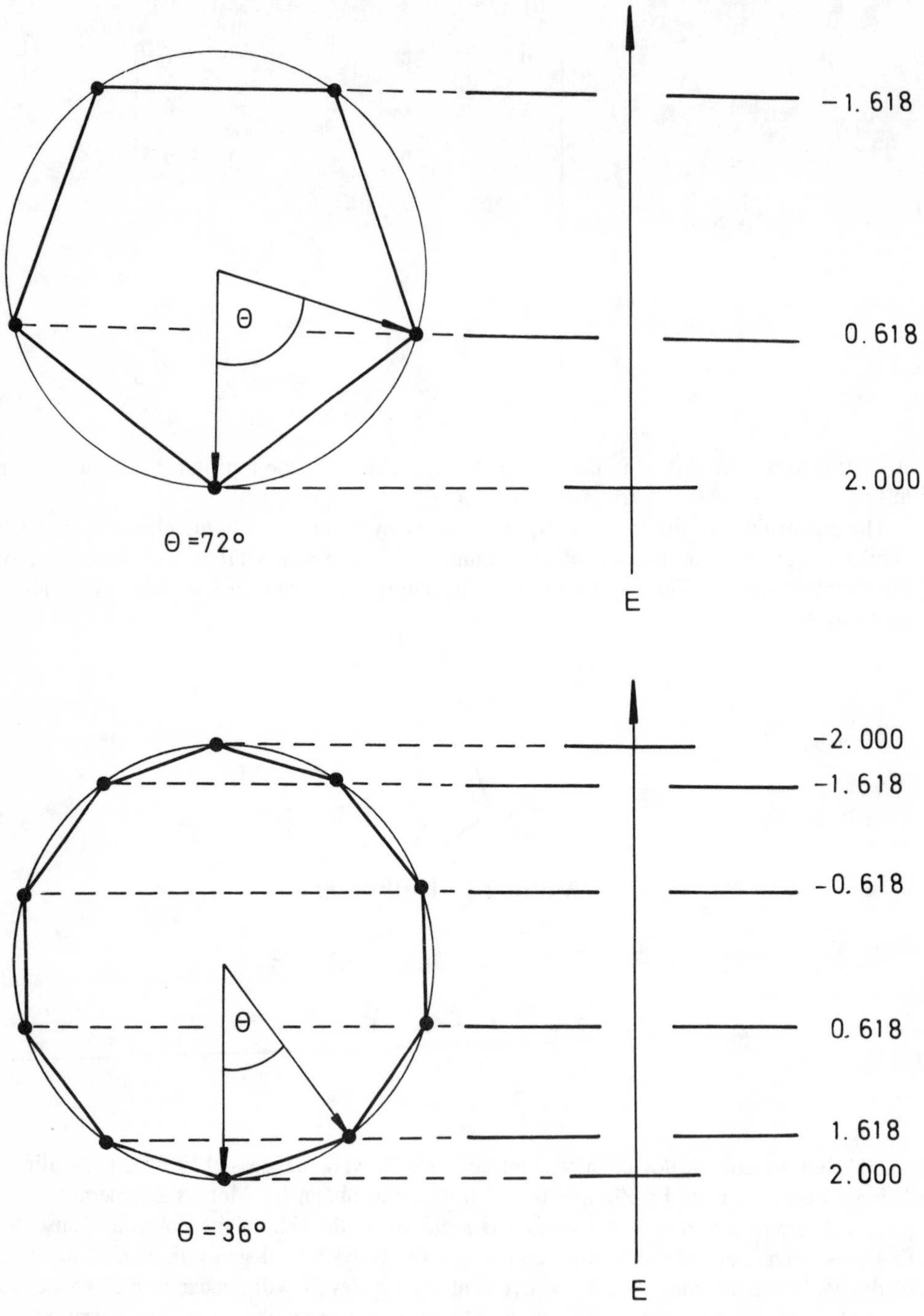

FIGURE 3. The Frost-Musulin projection diagrams for the eigenvalues of [5]- and [10]-annulenes.

IX. CLASSIFICATION SCHEME FOR MONOCYCLIC SYSTEMS

In Table 1 we report the polynomials and spectra of Hückel and Möbius [N]-annulenes from N = 3 up to N = 8.

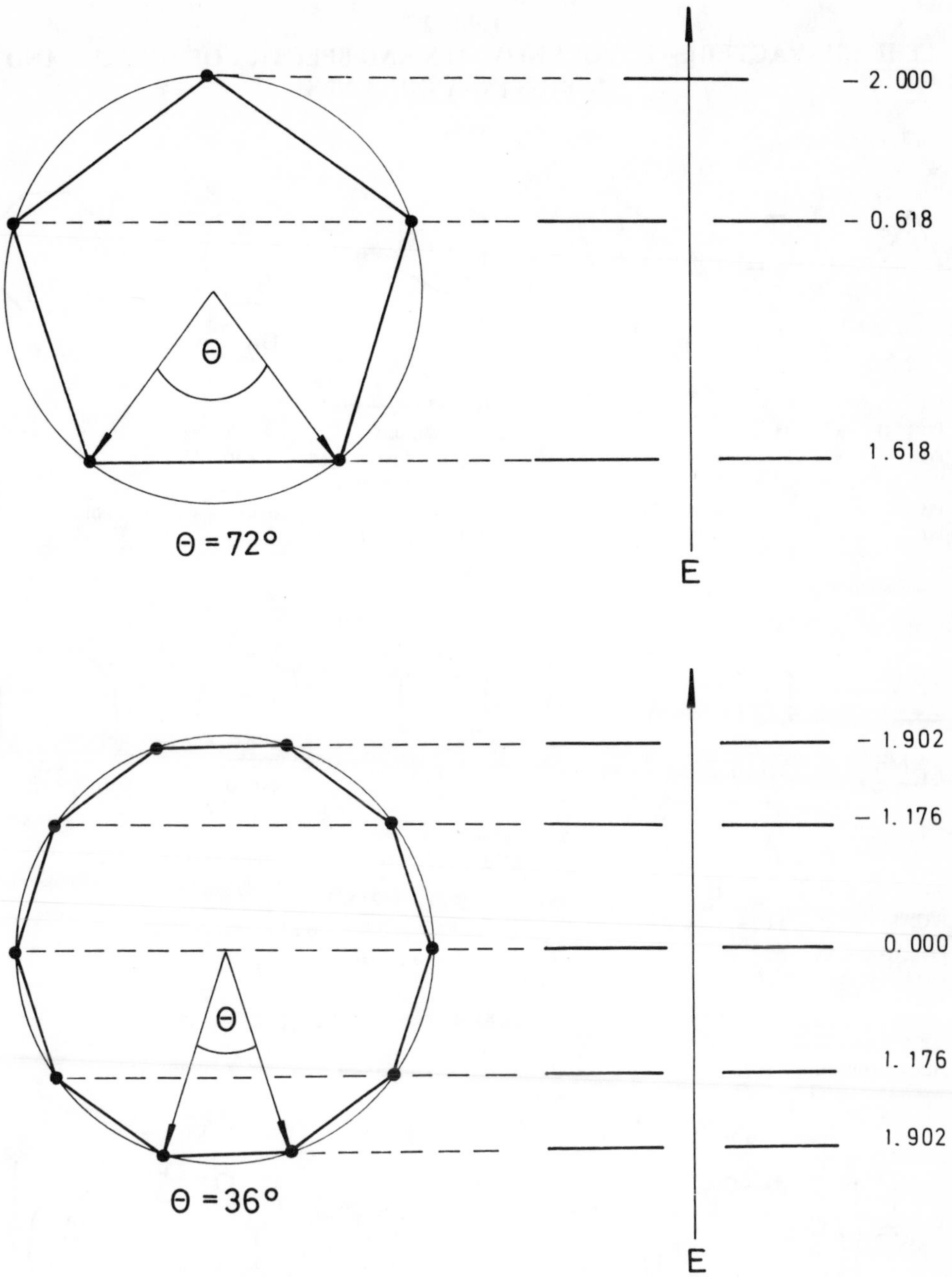

FIGURE 4. The Möbius-Zimmerman projection diagrams for the eigenvalues of Möbius [5]- and [10]-annulenes.

The inspection of the results in Table 1 shows that the *alternant* Hückel and Möbius annulenes differ only in the *value* of the a_N coefficient, while the *nonalternant* systems differ only in the *sign* of the a_N coefficient. This observation is not surprising because the monocyclic systems contain cycles only in the S_N set of Sachs graphs. Let us construct Sachs graphs with N vertices for Hückel and Möbius annulenes:

Table 1
THE CHARACTERISTIC POLYNOMIALS AND SPECTRA OF HÜCKEL AND MÖBIUS [N]-ANNULENES

Hückel [N]-annulenes

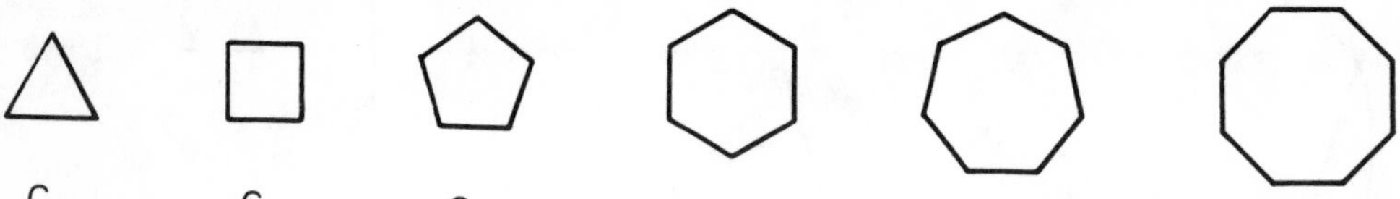

C_3 C_4 C_5 C_6 C_7 C_8

$P(C_3; x) = x^3 - 3x - 2$	{2.00, −1.00, −1.00}
$P(C_4; x) = x^4 - 4x^2$	{2.00, 0.00, 0.00, −2.00}
$P(C_5; x) = x^5 - 5x^3 + 5x - 2$	{2.0, 0.62, 0.62, −1.62, −1.62}
$P(C_6; x) = x^6 - 6x^4 + 9x^2 - 4$	{2.0, 1.0, 1.0, −1.0, −1.0, −2.0}
$P(C_7; x) = x^7 - 7x^5 + 14x^3 - 7x - 2$	{2.0, 1.25, 1.25, −0.45, −0.45, −1.80, −1.80}
$P(C_8; x) = x^8 - 8x^6 + 20x^4 - 16x^2$	{2.0, 1.41, 1.41, 0.0, 0.0, −1.41, −1.41, −2.0}

Möbius [N]-annulenes

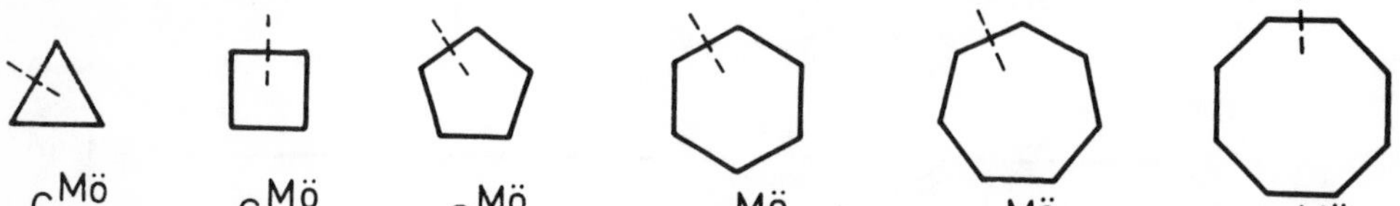

$C_3^{Mö}$ $C_4^{Mö}$ $C_5^{Mö}$ $C_6^{Mö}$ $C_7^{Mö}$ $C_8^{Mö}$

$P(C_3^{Mö}; x) = x^3 - 3x + 2$	{1.0, 1.0, −2.0}
$P(C_4^{Mö}; x) = x^4 - 4x^2 + 4$	$\{\sqrt{2}, \sqrt{2}, -\sqrt{2}, \sqrt{2}\}$
$P(C_5^{Mö}; x) = x^5 - 5x^3 + 5x + 2$	{1.62, 1.62, −0.62, −0.62, −2.0}
$P(C_6^{Mö}; x) = x^6 - 6x^4 + 9x^2$	$\{\sqrt{3}, \sqrt{3}, 0.0, 0.0, -\sqrt{3}, -\sqrt{3}\}$
$P(C_7^{Mö}; x) = x^7 - 7x^5 + 14x^3 - 7x + 2$	{1.8, 1.8, 0.45, 0.45, −1.25, −1.25, −2.0}
$P(C_8^{Mö}; x) = x^8 - 8x^6 + 20x^4 - 16x^2 + 4$	{1.85, 1.85, 0.77, 0.77, −0.77, −0.77, −1.85, −1.85}

Hückel annulenes

N = even

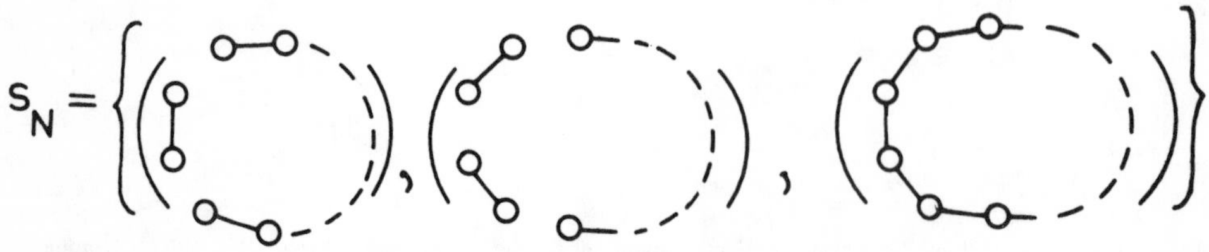

$$a_N = 2(-1)^{N/2}\, 2^0 + (-1)^1\, 2^1 = \begin{cases} 0 & \text{for } N = 4M \\ -4 & \text{for } N = 4m + 2 \end{cases} \quad (105)$$

N = odd

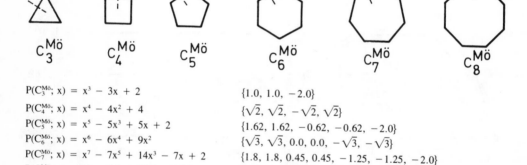

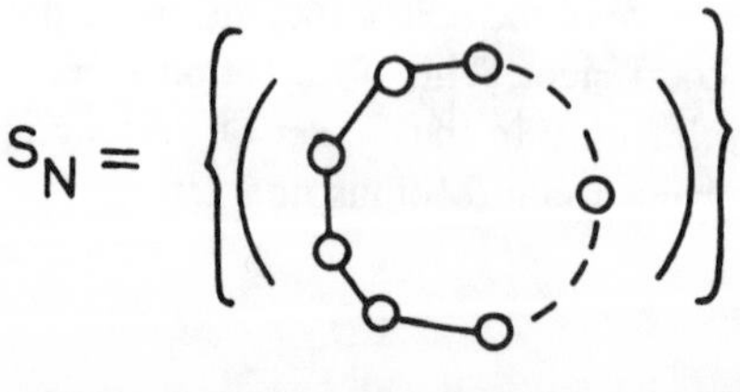

$$a_N = (-1)^1\ 2^1 = -2 \text{ for } N = 4m+1 \text{ or } N = 4m+3 \qquad (106)$$

Möbius annulenes
N = even

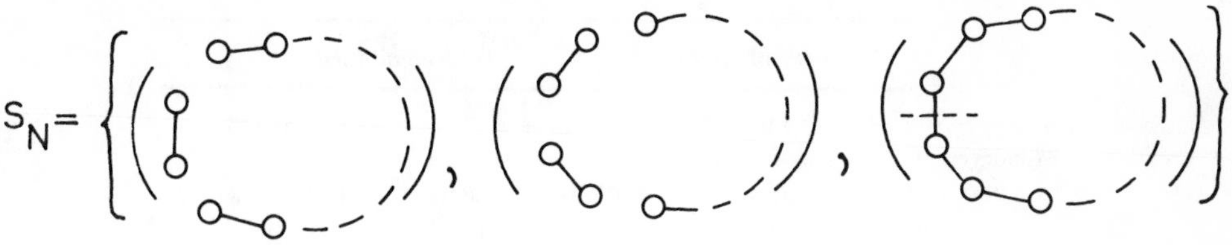

$$a_N = 2(-1)^{N/2+0}\ 2^0 + (-1)^{1+1}\ 2^1 = \begin{cases} 4 \text{ for } N = 4m \\ 0 \text{ for } N = 4m+2 \end{cases} \qquad (107)$$

N = odd

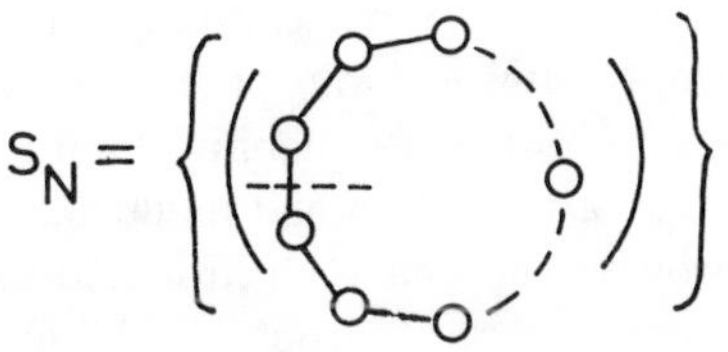

$$a_N = (-1)^{1+1}\ 2^1 = 2 \text{ for } N = 4m+1 \text{ or } N = 4m+3 \qquad (108)$$

The above results lead to the conclusion that the coefficient a_N may be used to establish a simple classification scheme for all monocyclic systems:

1. N = even

$$a_N = 0 \text{ for Hückel [4m]-annulenes and Möbius [4m+2]-annulenes}$$

$$a_N = \begin{cases} -4 \text{ Hückel [4m+2]-annulenes} \\ 4 \text{ Möbius [4m]-annulenes} \end{cases}$$

2. N = odd

$$a_N = \begin{cases} -2 \text{ Hückel [4m+1]- or [4m+3]-annulenes} \\ 2 \text{ Möbius [4m+1]- or [4m+3]-annulenes} \end{cases}$$

This scheme is a graph-theoretical justification of the *generalized* Hückel rule which embraces Hückel and Möbius annulenes and according to which Hückel [4m + 2]- and Möbius [4m]-annulenes should exhibit closed-shell stability and aromaticity, while Hückel [4m]- and Möbius [4m + 2]-annulenes should exhibit open-shell reactivity and antiaromaticity.[52,57-60] Hückel [4m + 1]- and [4m + 3]-annulenes and Möbius [4m + 1]- and [4m + 3]-annulenes are in between the [4m + 2]- and [4m]-systems. The stabilization or destabilization of these systems may attained by adding or substracting π-electrons. Hückel [4m + 1]-annulenes become quite *stable* species by *accepting* one π-electron and producing mono-anions, i.e., systems with (4m + 2) π-electrons and quite *unstable* species by *giving away* a π-electron

Table 2
GENERALIZED HÜCKEL RULE WHICH EMBRACES HÜCKEL AND MÖBIUS CYCLIC SYSTEMS

	Cycles[a]					
	Even-membered		Odd-membered			
	4m	4m + 2	4m + 1		4m + 3	
Monocyclic system			Cations	Anions	Cations	Anions
Hückel	−	+	−	+	+	−
Möbius	+	−	+	−	−	+

[a] (+) Denotes stable, aromatic species; (−) denotes unstable, antiaromatic species.

and producing mono-cations, i.e., monocyclic systems with (4m) π-electrons.[61,62] Hückel [4m + 3]-annulenes behave in the opposite way. They stabilize by giving away a π-electron and producing mono-cations, i.e., (4m + 2) π-electron systems, and destabilize by accepting one π-electron in their half-empty MOs and thus producing mono-anions, i.e., monocyclic systems with (4m) π-electrons.[61,62] Similarly, stabilization of Möbius [4m + 1]-annulenes may be achieved by giving away a π-electron and producing monocations. Destabilization of such a system is obtained by adding one π-electron and thus producing a mono-anion, i.e., a Möbius system with (4m + 2) π-electrons. A Möbius [4m + 3]-system stabilizes by receiving a π-electron and producing a mono-anion and destabilizes by giving away one π-electron and producing a mono-cation. All this may be conveniently tabulated as it is done in Table 2.

This is also the basis for the use of the Hückel-Möbius concept for qualitatively studying (poly)cyclic molecules and transition states of certain pericyclic reactions (i.e., electrocyclic closures of polyenes) by considering the topology of the corresponding transition states.[58,63-65]

X. TOTAL π-ELECTRON ENERGY

The total π-electron energy, E_π, is one of the most important pieces of information about the conjugated molecule which may be obtained from the HMO calculations since it can, in a proper way, be related to the thermodynamic stability of conjugated structures. Therefore, E_π is always positive and larger E_π means larger thermodynamic stability for molecules of the same size (the same N). A parametrization scheme, based on thermodynamical data, by Schaad and Hess,[66] has produced agreement with experiment of the same degree of quantitative accuracy as the much more sophisticated SCF MO procedure developed by Dewar.[67] In addition, Schaad and Hess[66] have shown that in many instances E_π follows linearly the total (thermodynamically measurable) energy of the conjugated compound. The physical reasons leading to an answer how it is possible that a model as simple as the Hückel model can give not only qualitative, but sometimes also fair quantitative agreement with the experimental findings are not well-understood at present. We accept the ''empirical'' assumption that the Hückel theory works and that it is a rather useful ''pencil and paper'' method for everyday use in the field of the chemistry of conjugated structures.

In this section, we will discuss how E_π depends on the structure of a conjugated molecule. The early work on this problem arose from the necessity to obtain numerical values of E_π in the times before the availability of electronic computers for quantum chemical problems. The foremost work in this precomputer era was carried out by Coulson.[39,68-72] Later, when computer facilities were readily available, the emphasis was to uncover the regularities relating E_π to selected topological invariants of the molecule.[73-80] The explicit or implicit use of chemical graph theory played a significant role in many of these investigations.

A. Fundamental Identity for E_π

Let us define two quantities: E_+ as the sum of the positive elements (bonding energy levels) N_+ of the Hückel spectrum (the Hückel MO energies), and E_- as the sum of the negative elements N_- of the Hückel spectrum, respectively,

$$E_+ = \sum_{i=1}^{N+} x_i \quad (109)$$

$$E_- = \sum_{i=N_+ + N_0 + 1}^{N} x_i \quad (110)$$

The relationship between E_+ and E_- is defined as follows,

$$E_+ + E_- = 0 \quad (111)$$

$$E_+ - E_- = \sum_{i=1}^{N} |x_i| \quad (112)$$

where $|x_i|$ is the absolute value of x_i. Since all the N_+ (bonding) levels in the molecule are ordinarily doubly occupied and all N_- (antibonding) levels are unoccupied, the total π-electron energy may be also given as,

$$E_\pi = 2\,E_+ \quad (113)$$

By means of relation (111), Equation (113) becomes,

$$E_\pi = E_+ - E_- \quad (114)$$

Coupling of (114) and (112) produces the identity,

$$E_\pi = \sum_{i=1}^{N} |x_i| \quad (115)$$

This identity is valid for alternant hydrocarbons because of the pairing theorem. In fact, the identity (115) holds if,

$$N_+ = N_- \quad (116)$$

However, if,

$$N_+ \neq N_- \quad (117)$$

the identity (115) is fulfilled approximately. This may be shown in the following way. If $N_+ \neq N_-$ and if there is a filled antibonding level $x_{N/2} < 0$, then,

$$E_\pi = \sum_{i=1}^{N} |x_i| + 2x_{N/2} \quad (118)$$

If,

$$\sum_{i=1}^{N} |x_i| >> 2\, x_{N/2} \qquad (119)$$

relation (115) holds as a good approximation. Similarly, if there is present an empty bonding energy level,

$$x_{N/2+1} > 0 \qquad (120)$$

then

$$E_\pi = \sum_{i=1}^{N} |x_i| - 2x_{N/2+1} \qquad (121)$$

Again if,

$$\sum_{i=1}^{N} |x_i| >> 2x_{N/2+1} \qquad (122)$$

relation (115) holds approximately. Thus, the relation (115) is fulfilled either exactly or as a very good approximation for all conjugated molecules (graphs). We refer to this result (115) as *the fundamental identity*[81] for E_π.

B. Relations between E_π, the Adjacency Matrix, and the Charge Density-Bond Order Matrix

Diagonalization of the adjacency matrix,

$$\mathbf{C\,A} = \mathbf{X\,C} \qquad (123)$$

produces the diagonal matrix **X**,

$$\mathbf{X} = \mathrm{diag}\,(x_1, x_2, \ldots, x_N) \qquad (124)$$

Let the function $f(x)$ be defined for all x_j, $j = 1, 2, \ldots, N$. Let also,

$$f(x) = \mathrm{diag}\,[f(x_1), f(x_2), \ldots, f(x_N)] \qquad (125)$$

Then, by definition, the matrix function is given by,

$$\mathbf{C}\, f\,(\mathbf{A}) = f(\mathbf{X})\,\mathbf{C} \qquad (126)$$

or

$$f(\mathbf{A}) = \mathbf{C}^{\dagger}\, f(\mathbf{X})\,\mathbf{C} \qquad (127)$$

In the scalar form Equation (127) becomes,

$$[f(\mathbf{A})]_{rs} = \sum_{i=1}^{N} f(x_i)\, c_{ir}\, c_{is} \qquad (128)$$

The above relation enables one to find relations between E_π, $\mathbf{A}(G)$, and $\mathbf{P}^C$. If a formal function $g = g(x)$ is introduced such that $g(x_i) = g_i$, the following relationship is obtained from (128),

$$\mathbf{P}^C = g(\mathbf{A}) \qquad (129)$$

In the case of filled bonding and empty antibonding molecular orbitals,

$$x_i\, g(x_i) = x_i + |x_i| \tag{130}$$

from which it immediately follows,

$$\mathbf{A}\,\mathbf{P}^{C} = \mathbf{A} + |\mathbf{A}| \tag{131}$$

$|\mathbf{A}|$ denotes the absolute value of the adjacency matrix **A** defined as,

$$|\mathbf{A}| = \mathbf{C}^{\dagger}\,\mathrm{diag}\,(|x_1|, |x_2|, \ldots, |x_N|)\,\mathbf{C} \tag{131a}$$

Since the trace of the matrix is not changed by unitary transformation,

$$\mathrm{Tr}\,|\mathbf{A}| = \mathrm{Tr}\,[\mathrm{diag}\,(|x_1|, |x_2|, \ldots, |x_N|)] = \sum_{i=1}^{N} |x_i| \tag{132}$$

Use of the fundamental identity leads to relation between E_π and **A**,

$$E_\Pi = \mathrm{Tr}\,|\mathbf{A}| \tag{133}$$

Taking into account that Tr $\mathbf{A} = 0$, and combining (133) and (131), we obtain the relation between E_π, **A**, and $\mathbf{P}_C$,

$$E_\Pi = \mathrm{Tr}\,\mathbf{A}\,\mathbf{P}^{C} \tag{134}$$

Both relations (133) and (134) were first obtained by Ruedenberg[82,83] although they were known to Hall[42] for alternants.

C. The McClelland Formula for E_π

The total π-electron energy is a bounded quantity. The bounds may be given, by using the Frobenius theorem, as follows,

$$0 \leqq E_\pi \leqq N\,D_{max} \tag{135}$$

McClelland[73] was the first to show that much better bounds for E_π can be deduced. Let,

$$F = 2M - N\,(\det \mathbf{A})^{2/N} \tag{136}$$

Note that $F > 0$. Then the McClelland inequalities are,

$$0 \leqq 2NM - E_\pi^2 \leqq (N-1)\,F \tag{137}$$

The above was later improved,[84]

$$F \leqq 2NM - E_\pi^2 \leqq (N-1)\,F \tag{138}$$

Table 3
HMO ENERGIES AND THE CORRESPONDING MCCLELLAND VALUES OF SOME CONJUGATED MOLECULES

Molecule	HMO energy[a]	McClelland energy[b]
Benzene	8.000	7.806
Fulvene	7.466	7.806
Heptafulvene	9.994	10.409
Styrene	10.424	10.409
o-Xylylene	9.954	10.409
m-Xylylne	9.431	10.409
p-Xylylene	9.925	10.409
Biphenyl	16.383	16.250
Azulene	13.364	13.646
Pentalene	10.456	11.040
Naphthalene	13.683	13.646
Anthracene	19.314	19.472
Phenanthrene	19.448	19.472
Pyrene	22.505	22.685
Naphthacene	24.931	25.296

[a] Normalized ($\alpha = 0$, $\beta = 1$) Hückel values are taken from C. A. Coulson and A. Streitwieser, Jr., *Dictionary of π-Electron Calculations*, W. H. Freeman, San Francisco, 1965.
[b] Normalized McClelland values are obtained from the formula $E_\pi = 0.92\ (2{\cdot}M{\cdot}N)^1\nabla^2$.

Relation (138) holds for all graphs. For bipartite graphs it slightly alters,

$$F \leqq 2\,N\,M - E_\pi^2 \leqq (N - 2)\,F \tag{139}$$

The importance of inequalities (137) to (139) is in that they reveal the most important topological factors in determining E_π: (1) the number of atoms N, (2) the number of bonds M, and (3) the determinant of the adjacency matrix *det A*.

McClelland[73] derived an approximate formula for E_π,

$$E_\Pi \approx a(2\,M\,N)^{1/2} \tag{140}$$

where the value of the constant *a* can be obtained by least-squares fitting. The optimal value $a = 0.92$ has reproduced rather closely E_π of various conjugated molecules; the difference between the exact and approximate values being only a few percent (see Table 3).

This numerical work clearly shows that the gross part of E_π is determined solely by molecular size. Since molecular size is defined by only two topological parameters — the number of atoms and bonds — all other topological factors play, therefore, a seemingly marginal role. However, in chemical applications we are interested not only in energies, but in energy differences also. Thus, the problem of a few percent of E_π is essential for chemistry. Moreover, the formula (140) fails to differentiate isomeric structures, because isomers have identical values of M and N quantities. Thus, it cannot be used, for example, for stability predictions.[77]

It can be also shown that E_π is proportional to the number of rings in a molecule. The number of rings R in a molecule is given by,

$$R = M - N + 1 \tag{141}$$

By susbtituting R + N − 1 for M in (140) we obtain,

$$E_\pi \approx a\,[2N\,(R + N - 1)]^{1/2} \tag{142}$$

For sufficiently large molecules N >> R − 1 and E_π should be almost linearly proportional to the number of vertices. Such linear dependences are observed in various homologous series, i.e., annulenes,[85] radialenes,[86] polyacenes,[87,88] etc. A practical result of Equation (142) is that one cannot compare conjugated molecules with the same N, but different R. For example, azulene should not be compared with [10]-annulene because according to Equation (142) azulene should necessarily have a greater E_π.

REFERENCES

1. **Hückel, E.,** *Z. Phys.*, 60, 204, 1931.
2. **Hückel, E.,** *Z. Phys.*, 72, 310, 1932.
3. **Hückel, E.,** *Z. Phys.*, 76, 628, 1932.
4. **Gutman, I. and Trinajstić, N.,** *Topics Curr. Chem.*, 42, 49, 1973.
5. **Coulson, C. A., O'Leary, B., and Mallion, R. B.,** *Hückel Theory for Organic Chemists,* Academic Press, London, 1978.
6. **Streitwieser, A., Jr.,** *Molecular Orbital Theory for Organic Chemists,* John Wiley & Sons, New York, 1961.
7. **Salem, L.,** *The Molecular Orbital Theory of Conjugated Systems,* Benjamin, New York, 1966.
8. **Heilbronner, E. and Bock, H.,** *The HMO Model and Its Application,* John Wiley & Sons, London, 1976.
9. **Bloch, F.,** *Z. Phys.*, 52, 555, 1929, 61, 206, 1930.
10. **Coulson, C. A. and Streitwieser, A., Jr.,** *Dictionary of π-Electron Calculations,* W. H. Freeman, San Francisco, 1965.
11. **Ruedenberg, K.,** *J. Chem. Phys.*, 22, 1878, 1954.
12. **Günthard, H. H. and Primas, H.,** *Helv. Chim. Acta,* 39, 1645, 1956.
13. **Schmidtke, H.-H.,** *J. Chem. Phys.*, 45, 3920, 1966.
14. **Gutman, I. and Trinajstić, N.,** *Croat. Chem. Acta,* 47, 507, 1975; Math. Chem. (Mülheim/Ruhr), 1, 71, 1975.
15. **Trinajstić, N.,** in *Semiempirical Methods of Electronic Structure Calculation. Part A: Techniques,* Vol. 7 Segal, G. A., Ed., Plenum Press, New York, 1977, 1.
16. **Ruedenberg, K.,** *J. Chem. Phys.*, 34, 1861, 1961.
17. **Cvetković, D., Gutman, I., and Trinajstić, N.,** *Chem. Phys. Lett.*, 29, 65, 1974.
18. **Longuet-Higgins, H. C.,** *J. Chem. Phys.*, 18, 265, 1950.
19. **Jahn, G. A. and Teller, E.,** *Proc. R. Soc. London Ser. A,* 161, 220, 1937.
20. **Clar, E., Kemp, W., and Stewart, D. C.,** *Tetrahedron,* 3, 36, 1958.
21. **Lin, C. Y. and Krantz, A.,** *J. Chem. Soc., Chem. Commun.*, 1111, 1972.
22. **Živković, T.,** *Croat. Chem. Acta,* 44, 351, 1972.
23. **Cvetković, D., Gutman, I., and Trinajstić, N.,** *J. Mol. Struct.*, 28, 289, 1975; **Cvetković, D., Gutman, I., and Trinajstić, N.,** *Croat. Chem. Acta,* 44, 365, 1972.
24. **Gutman, I., Trinajstić, N., and Živković, T.,** *Tetrahedron,* 29, 3349, 1973.
25. **Gutman, I.,** *Chem. Phys. Lett.*, 26, 85, 1974.
26. **Kurosh, A. G.,** *Higher Algebra,* Mir, Moscow, 1980, 247, third printing.
27. **Gutman, I. and Trinajstić, N.,** *Naturwissenschaften,* 60, 475, 1973.
28. **Lloyd, D.,** *Carbocyclic Non-Benzenoid Aromatic Compounds,* Elsevier, Amsterdam, 1966.
29. **Katz, T. J. and Rosenberg, M.,** *J. Am. Chem. Soc.*, 84, 865, 1962.
30. **Coulson, C. A.,** *Proc. R. Soc. London, Ser. A,* 169, 413, 1939.
31. **Mulliken, R. S.,** *J. Chem. Phys.*, 23, 1841, 1955.
32. **Ham, N. S. and Ruedenberg, K.,** *J. Chem. Phys.*, 29, 1215, 1958.
33. **Coulson, C. A. and Golebiewski, A.,** *Proc. Phys. Soc. (London),* 78, 1310, 1961.
34. **Pauling, L., Brockway, L. O., and Beach, J. Y.,** *J. Am. Chem. Soc.*, 57, 2705, 1935.
35. **Herndon, W. C.,** *J. Am. Chem. Soc.*, 96, 7605, 1974.

36. **Randić, M.,** *Croat. Chem. Acta,* 47, 71, 1975.
37. **Herndon, W. C. and Párkányi, C.,** *J. Chem. Educ.,* 53, 689, 1976.
38. **Ham, N. S.,** *J. Chem. Phys.,* 29, 1229, 1958.
39. **Coulson, C. A. and Longuet-Higgins, H. C.,** *Proc. R. Soc. London, Ser. A,* 192, 16, 1947.
40. **Coulson, C. A. and Rushbrooke, G. S.,** *Proc. Cambridge Phil. Soc.,* 36, 193, 1940.
41. **Moffitt, W.,** *J. Chem. Phys.,* 26, 424, 1957.
42. **Hall G. G.,** *Proc. R. Soc. London, Ser. A,* 229, 251, 1955.
43. **McLachlan, A. D.,** *Mol. Phys.,* 2, 271, 1959.
44. **Gołebiewski, A.,** *Acta Phys. Polonica,* 23, 235, 1963.
45. **Koutecký, J.,** *J. Chem. Phys.,* 44, 3702, 1966.
46. **Graovac, A., Gutman, I., Trinajstić, N., and Živković, T.,** *Theor. Chim. Acta,* 26, 67, 1972.
47. **Mallion, R. B., Schwenk, A. J., and Trinajstić, N.,** in *Recent Advances in Graph Theory,* Fiedler, M., Ed., Academia, Prague, 1975, 345.
48. **McClellan, A. L.,** *Tables of Experimental Dipole Moments,* W. H. Freeman, San Francisco, 1963.
49. **Kirsanov, B. P. and Basilevsky, M. V.,** *Zh. Strukt. Khim.,* 5, 99, 1964.
50. **Rutherford, D. E.,** *Proc. R. Soc. Edinburgh, Ser. A,* 62, 229, 1947; see also **Lennard-Jones, J. E.,** Proc. R. Soc. London, Ser. A, 158, 280, 1937.
51. **Frost, A. A. and Musulin, B.,** *J. Chem. Phys.,* 21, 572, 1953.
52. **Sondheimer, F.,** *Acc. Chem. Res.,* 5, 81, 1972.
53. **Polansky, O. E.,** *Monatsh. Chem.,* 91, 916, 1960.
54. **Marcus, H. and Minc, H.,** *A Survey of Matrix Theory and Matrix Inequalities,* Allyn & Bacon, Boston, 1964, 66.
55. **Mirsky, L.,** *Introduction to Linear Algebra,* Oxford University Press, Oxford, 1955, 36.
56. **Graovac, A. and Trinajstić, N.,** *Croat. Chem. Acta,* 47, 95, 1975.
57. **Heilbronner, E.,** *Tetrahedron Lett.,* 1923, 1964.
58. **Zimmerman, H. E.,** *J. Am. Chem. Soc.,* 88, 1564, 1971; *Acc. Chem. Res.,* 4, 272, 1971.
59. **Mason, S.,** *Nature (London),* 205, 495, 1965.
60. **Sondheimer, F.,** *Pure Appl. Chem.,* 7, 363, 1963; *Proc. Robert A. Welch Found. Conf. Chem. Res.,* 12, 125, 1968.
61. **Garratt, P. J. and Sargent, M. V.,** in *Advances in Organic Chemistry,* Taylor, E. C. and Wynberg, H., Eds., John Wiley & Sons, New York, 1969, 1.
62. **Cresp, T. M. and Sargent, M. V.,** in *Essays in Chemistry,* Vol. 4, Bradley, J. N., Gillard, R. D., and Hudson, R. F., Eds., Academic Press, London, 1972.
63. **Shen, K.-W.,** *J. Chem. Educ.,* 50, 238, 1971.
64. **Dewar, M. J. S.,** *Angew. Chem. Int. Ed.,* 10, 761, 1971.
65. **Smith, W. B.,** *Molecular Orbital Methods in Organic Chemistry: HMO and PMO,* Marcel Dekker, New York, 1974.
66. **Schaad, L. J. and Hess, B. A., Jr.,** *J. Am. Chem. Soc.,* 94, 3068, 1972.
67. **Dewar, M. J. S.,** *The Molecular Orbital Theory of Organic Chemistry,* McGraw-Hill, New York, 1969.
68. **Coulson, C. A.,** *Proc. Cambridge Phil. Soc.,* 36, 201, 1940.
69. **Coulson, C. A. and Longuet-Higgins, H. C.,** *Proc. R. Soc. London, Ser. A,* 191, 39, 1947.
70. **Coulson, C. A. and Jacobs, J.,** *J. Chem. Soc.,* 2805, 1949.
71. **Coulson, C. A.,** *Proc. Cambridge Phil. Soc.,* 46, 202, 1950.
72. **Coulson, C. A.,** *J. Chem. Soc.,* 3111, 1954.
73. **McClelland, B. J.,** *J. Chem. Phys.,* 54, 640, 1971.
74. **Gutman, I. and Trinajstić, N.,** *Chem. Phys. Lett.,* 17, 535, 1972.
75. **Gutman, I., Trinajstić, N., and Živković, T.,** *Chem. Phys. Lett.,* 14, 342, 1972.
76. **Hall, G. G.,** *Int. J. Math. Educ. Sci. Technol.,* 4, 233, 1973.
77. **Gutman, I., Milun, M., and Trinajstić, N.,** *J. Chem. Phys.,* 59, 2772, 1973.
78. **Gutman, J.,** *Chem. Phys.,* 66, 1652, 1977.
79. **Hall, G. G.,** *Mol. Phys.,* 33, 551, 1977.
80. **Gutman, I.,** *Theor. Chim. Acta,* 45, 79, 1977.
81. **Graovac, A., Gutman, I., and Trinajstić, N.,** *Topological Approach to the Chemistry of Conjugated Molecules, Lecture Notes in Chemistry,* Vol. 4, Springer-Verlag, Berlin, 1977.
82. **Ruedenberg, K.,** *J. Chem. Phys.,* 29, 1232, 1958.
83. **Ruedenberg, K.,** *J. Chem. Phys.,* 34, 1884, 1961.
84. **Gutman, I.,** *Chem. Phys. Lett.,* 24, 283, 1974.
85. **Gutman, I., Milun, M., and Trinajstić, N.,** *Croat. Chem. Acta,* 44, 207, 1972.
86. **Gutman, I., Trinajstić, N., and Živković, T.,** *Croat. Chem. Acta,* 44, 501, 1972.
87. **Heilbronner, E.,** *Helv. Chim. Acta,* 37, 921, 1954.
88. **England, W. and Ruedenberg, K.,** *J. Am. Chem. Soc.,* 95, 8769, 1973.

Chapter 7

ISOSPECTRAL MOLECULES

At one time,[1-4] it was supposed that the characteristic polynomial and its spectrum might be a unique property of a graph. Therefore graphs with identical spectra would be identical or isomorphic. However, the counterexamples have shown that two or more graphs which are nonisomorphic may have identical spectra.[5-9] Nonidentical graphs with identical spectra are called *isospectral*[10] or *cospectral*[11] graphs. A classical example[10,12] is provided by molecular graphs **1** and **2** corresponding to 1,4-divinylbenzene and 2-phenylbutadiene.*

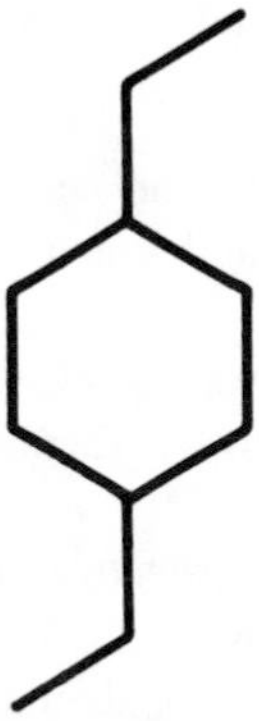

1

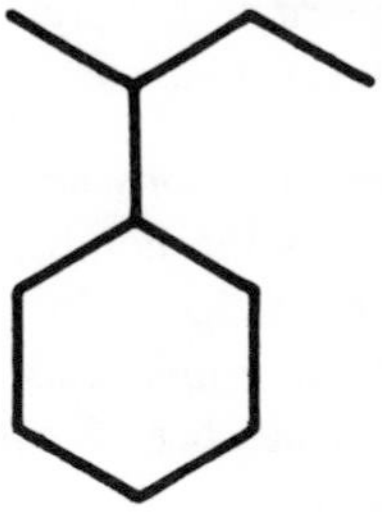

2

Their polynomials are identical: $P(G_1; x) = P(G_2; x) = x^{10} - 10x^8 + 33x^6 - 44x^4 + 24x^2 - 4$ and, consequently, their spectra are also identical: spectrum of **1** = spectrum of **2** = (± 2.214, ± 1.675, ± 1.000, ± 1.000, ± 0.539). This kind of molecules are named *the isospectral molecules.*[10,12-16]

The isospectrality concept has found use in physics,[17] applied mathematics,[9,18-22] and in recent years, chemistry.[10,12-16] Although the Hückel values of the isospectral molecules were known for 20 years, the reasons for the identity of their Hückel energy levels have been found only recently by means of graph theory. Here we will show how the occurrence of the isospectral molecules is strongly related to the topology of their molecular skeletons.

There are three problems we wish to address concerning isospectral molecules:

1. Given two nonidentical molecules, how can one tell if they are isospectral?
2. What are the available methods for generating isosopectral molecules?
3. Given a particular molecule, is it possible to determine whether one or more isospectral mates exist and if so can one derive the isospectral partners?

I. DETERMINATION OF ISOSPECTRALITY

The first problem is the easiest to solve. Several methods exist for determining isospectrality of a pair of molecules. The obvious brute-force way is to calculate the eigenvalues of the adjacency matrix of each molecule and compare eigenvalues. If the two sets of eigenvalues match exactly, then the two molecules are isospectral. Another way is to work out the characteristic polynomial of the adjacency matrix of each molecule. If the coefficients

* Dr. Tomislav Živković (Zagreb) appears to be the first to observe that 1,4-divinylbenzene and 2-phenylbutadiene are isospectral. This was reported at the Quantum Chemistry School, Repino near Leningrad, U.S.S.R., December, 1973.

of comparable terms in the characteristic polynomials are identical term by term, then the molecules are isospectral.

The coefficients of the characteristic polynomial can be determined individually by the Sachs procedure[23,24] as described in Chapter 5.

The existence of Sachs expressions for constructing the individual coefficients suggests that one could make a term-by-term calculation and comparison of characteristic polynomial coefficients for a pair of graphs: the first nonidentical pair of comparable coefficients that arises would demonstrate nonisospectrality and further comparisons could be discontinued. The simplicity of the expressions for the coefficients a_2 and a_3 (Equations 57 and 58, Chapter 5) leads to the following simple counting rules that can quickly reveal basic differences between two graphs that prevent them being isospectral. These conditions are necessary but not sufficient for isospectrality.

1. *Isospectral graphs must have the same number of vertices.* The number of vertices determines the numbers of eigenvalues in the spectrum and graphs with different numbers of eigenvalues cannot be isospectral.
2. *Isospectral graphs must have the same number of edges.* Since $a_2 = -e$, graphs with different number of edges will have different values of a_2 and therefore cannot be isospectral.
3. *An acyclic graph and a graph containing rings cannot be isospectral.* To form a ring requires an additional edge beyond those present in the acyclic graph. Therefore, cyclic graphs necessarily have more edges and different coefficients a_n than acyclic graphs. This condition holds only for the connected graphs.
4. *Isospectral graphs without fused rings must have the same number of rings.* The same number of nonfused rings assures that the graphs will have the same number of edges and the same a_2. For graphs with fused rings the situation is more complicated because the number of fused rings is not related to the number of edges in a linear way.
5. *Isospectral graphs must have the same number of three-membered rings.* Since $a_3 = -2\ C_3$, graphs with different numbers of three-membered rings will have different values of a_3 and therefore cannot be isospectral.
6. *Graphs containing only even-membered rings cannot be isospectral with those containing any odd-membered rings.* The Sachs expressions for all the odd coefficients a_{2k+1} are zero for graphs containing no odd-membered rings. For a graph containing odd-membered rings at least one a_{2k+1} must be nonzero. Another justification for this rule is based on the fact that the alternant systems contain no odd-membered rings[25] and for every eigenvalue there is another eigenvalue of equal magnitude but opposite sign (the pairing theorem).[26] Nonalternant systems must contain at least one odd-membered ring and their energy levels are not paired. Therefore, alternant and nonalternant systems cannot be isospectral.

Rules 1 to 6 are specific statements of the fact that isospectral molecules must be isomers.

Heilbronner[27] has shown how to factor the secular determinant or characteristic polynomial of a composite molecule C composed of two fragments A and B linked by a bond between atom *a* in A and atom *b* in B. For normal homo-atomic molecules the characteristic polynomial of the composite molecule C can be symbolically expressed as

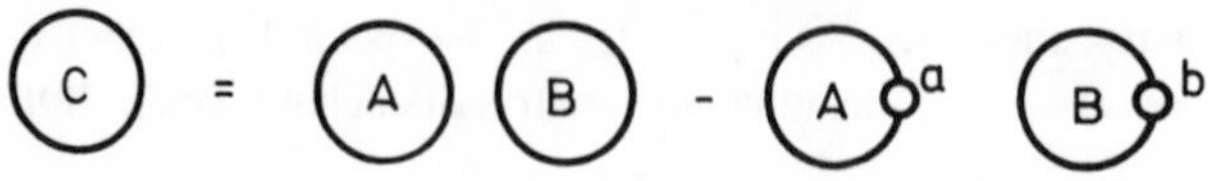

(1)

where  and  represent the characteristic polynomials of fragments A and B, P(A; x) and P(B; x), and 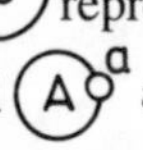 and 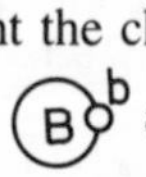 are the characteristic polynomials of fragments A and

B with atoms *a* and *b* removed, P(A-a; x) and P (B-b; x), respectively. Equation (1) is restated below in different notation,

$$P(C;x) = P(A;x)\,P(B;x) - P(A - a;x)\,P(B - b;x) \qquad (2)$$

Factorization of this type can reduce the characteristic polynomial to smaller pieces which either can be immediately seen to be identical or else the polynomials of the fragments can be easily worked out, perhaps by the Sachs procedure, and then compared for isospectrality. As an example, break up the acyclic graphs **3** and **4** across the bonds as indicated.

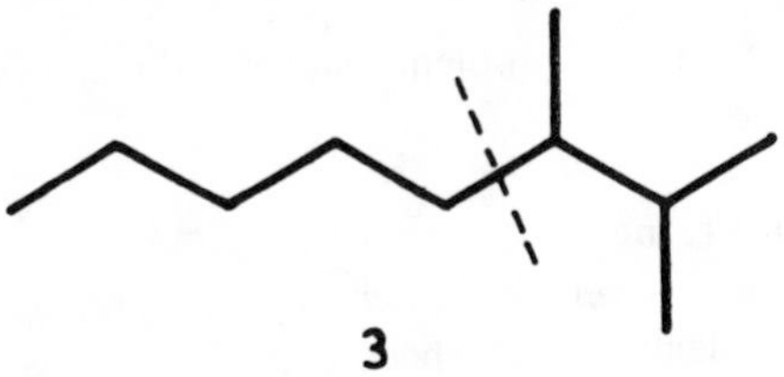

3

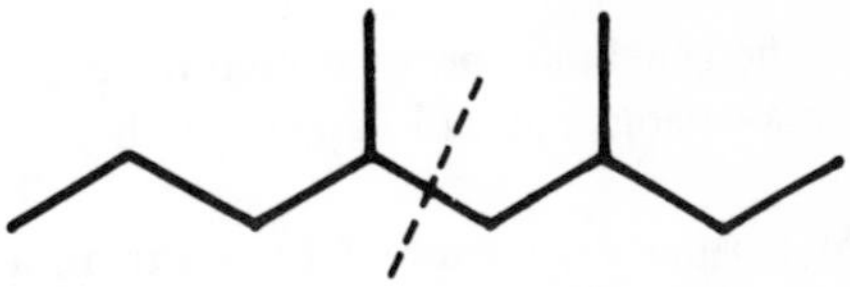

4

Applying the Heilbronner factorization formula for the characteristic polynomials we obtain,

$$P(3;x) = [L_5]\cdot$$  $$- [L_4]\cdot[L_3]\cdot x \qquad (3)$$

$$P(4;x) = [L_5]\cdot$$ 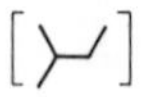 $$- [L_3]\cdot[L_4]\cdot x \qquad (4)$$

Where L_N is the characteristic polynomial of the linear graph (chain) with N vertices and 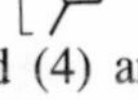 is the polynomial for the structure enclosed in brackets. The expressions (3) and (4) are identical hence the graphs 3 and 4 are *isospectral mates.*[9]

Randić[28] has proposed an interesting criterion for investigating whether two graphs are isospectral or not. He uses the fact that the characteristic polynomial is related to random walks.[29] (A *random walk* in a graph is a sequence of edges which can be continuously traversed, starting from any vertex and ending on any vertex, also permitting the use of the same edge several times.)[29] This relationship is reflected in the identity of the information given by the characteristic polynomial and spectral moments of a graph. The random walks considered here are *self-returning walks,* i.e., random walks starting and ending at the same vertex. Self-returning walks starting at a vertex *q* are given by the *q*-th diagonal elements of the adjacency matrix **A** raised to the powers *n:* $(\mathbf{A}^n)_{qq}, n = 1,2, \ldots, N$. For each *n,* the spectral moment of a graph is the trace of the matrix $\mathbf{A}^n$. Two graphs are isospectral if they have identical sequences of spectral moments, $n = 1,2, \ldots, N$.

As an example, we consider the Balaban-Harary isospectral pair:[6]

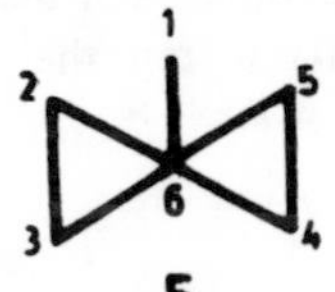

5

Graph 5

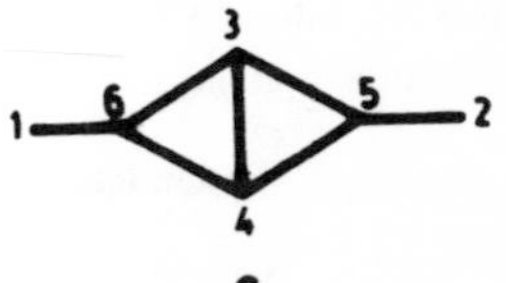

6

Graph 6

Numbering of atoms	Self-returning walks					
	n = 1	2	3	4	5	6
1	0	1	0	5	4	29
2,3,4,5	0	2	2	9	18	58
6	0	5	4	29	44	185
spectral moments:	0	14	12	70	120	446

Numbering of atoms	Self-returning walks					
	n = 1	2	3	4	5	6
1,2	0	1	0	3	2	15
3,4	0	3	4	17	36	115
5,6	0	3	2	15	22	93
spectral moments:	0	14	12	70	120	446

Graphs **5** and **6** have identical spectral moments and thus represent isospectral mates.

Another point of view is that of deciding whether identical spectra originate from identical graphs or nonidentical graphs. Randić has proposed a procedure that he claims can establish the identity of two apparently different graphs.[30,31] It is based on a direct comparison of the adjacency matrices and hence he has developed a procedure for setting up these matrices uniquely.

II. CONSTRUCTION OF ISOSPECTRAL SYSTEMS

The generation of isospectral molecules is a more complex problem, but several methods are available and will be exposed here.

A. Isospectral Graphs Derived from a Common Frame

Consider a graph which contains one or more pairs of vertices which have the property that the removal of each vertex in turn produces two identical or isospectral graphs. Vertices having this property may be called *isospectral points*[13,14] or *active sites*.[12,15] A graph containing isospectral points may serve as the common *frame*[12] in a newly-constructed isospectral pair of graphs, derived from the common frame by attaching two different graphical fragments to the isospectral points in a reciprocal relationship.

The isospectral points in the styrene graph **7**, for example, are the vertices 2 and 6 in the graph below.

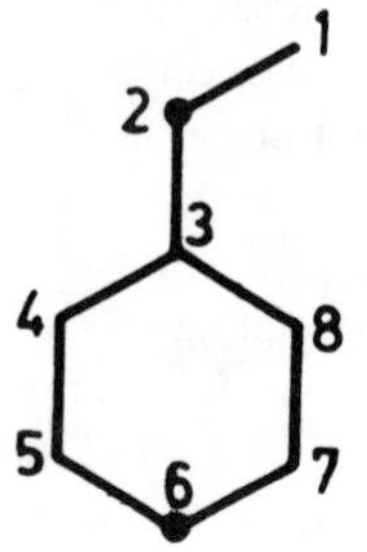

7

Randić[28] has produced a valuable criterion for detecting the isospectral points. The isospectral points always have identical sequences of atom self-returning walks. For the graph **7** the following list of self-returning walks for different atoms is derived from the powers of the adjacency matrix.

Atom label	Self-returning walk							
	n = 1	2	3	4	5	6	7	8
1	0	1	0	2	0	6	0	22
2	0	2	0	6	0	22	0	90
3	0	3	0	12	0	52	0	235
4,8	0	2	0	7	0	30	0	135
5,7	0	2	0	6	0	23	0	98
6	0	2	0	6	0	22	0	90

As expected, the positions 2 and 6 have identical sequences of self-returning walks.

An examination of their tabulated eigenvalues[32-34] confirms the fact that the graph resulting from the removal of the vertex 2, denoted by **7**-v_2, and the graph resulting from the removal of the vertex 6, denoted by **7**-v_6, are isospectral.

7 - V_2 7 - V_6

(The symbol " ≡ " is used here to signify an isospectral relationship.) Since 2 and 6 are isospectral points, two arbitrary graphical fragments A and B may be attached to 2 and 6 in a reciprocal manner (i.e., by an ordered replacement) to generate a new pair of isospectral graphs **8** and **9**.

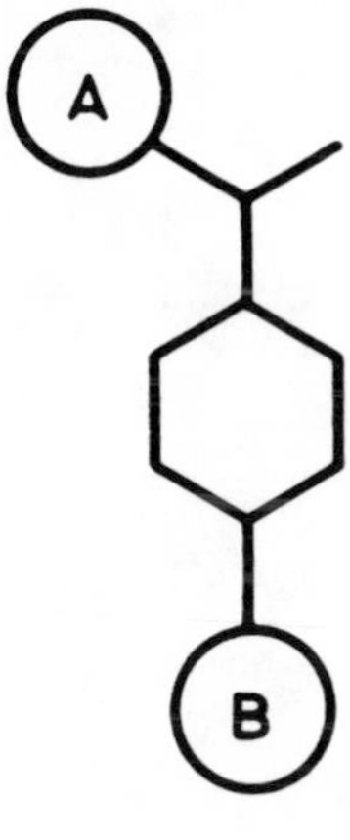

8

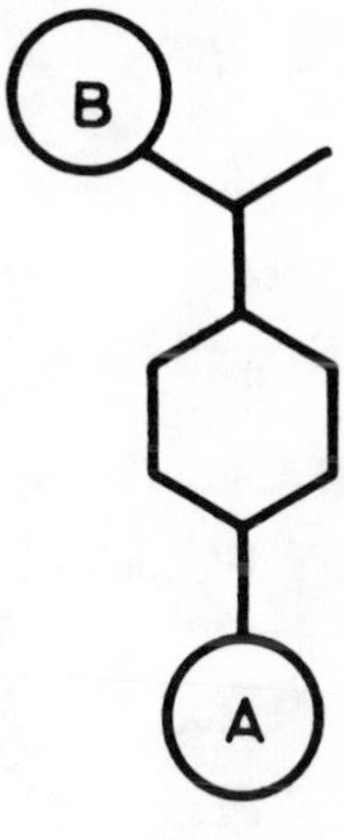

9

The graph **10,** shown below, contains a pair of vertices whose successive removal generates a pair of identical, rather than isospectral, graphs.

10

Since the vertices 2 and 5 are isospectral points in **10,** two isospectral graphs **11** and **12** may be constructed from **10** by the reciprocal attachment of two arbitrary fragments C and D at 2 and 5.

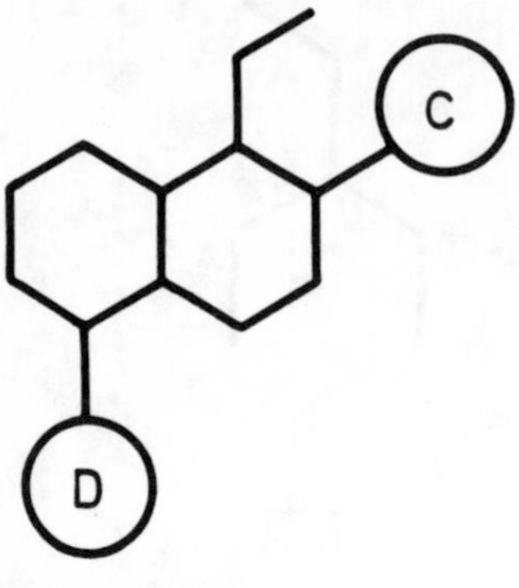

11

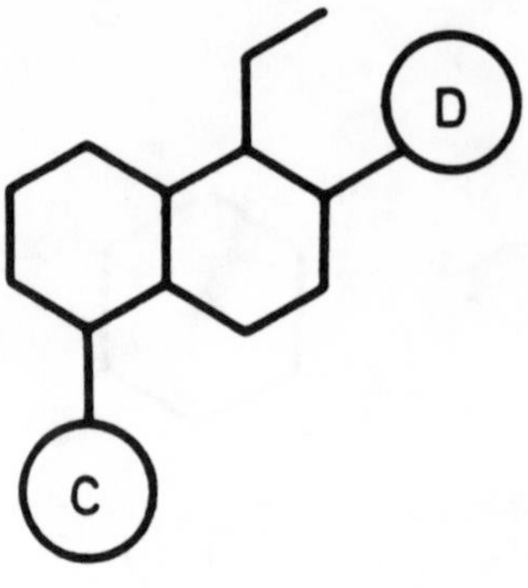

12

In many cases, the original graphical frame contains more than one pair of isospectral points. The graph **10,** for example, has as a second pair of isospectral points the vertices labeled 7 and 11 in the graph below.

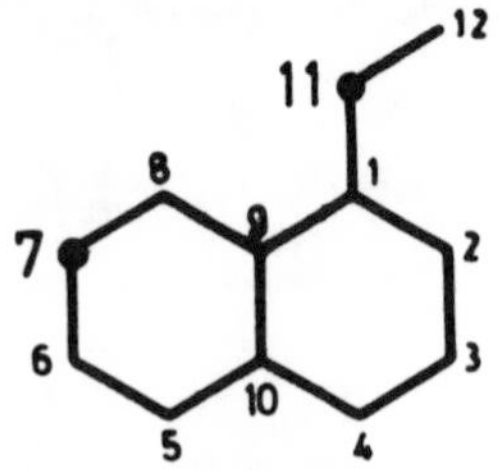

10

The successive deletion of 7 and 11 from **10** produces the two isospectral graphs **10**-v_7 and **10**-v_{11}.

10 - v_7 10 - v_{11}

It is thus possible to construct an isospectral quartet **13, 14, 15,** and **16** by first attaching two fragments A and B to **10** at sites 7 and 11, then attaching two additional fragments C and D at sites 2 and 5.

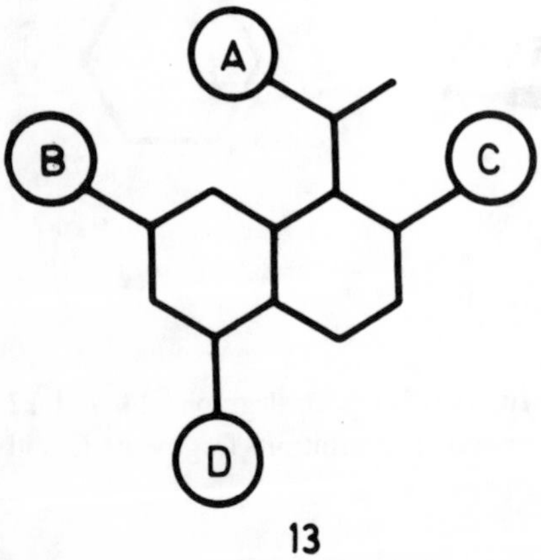

13

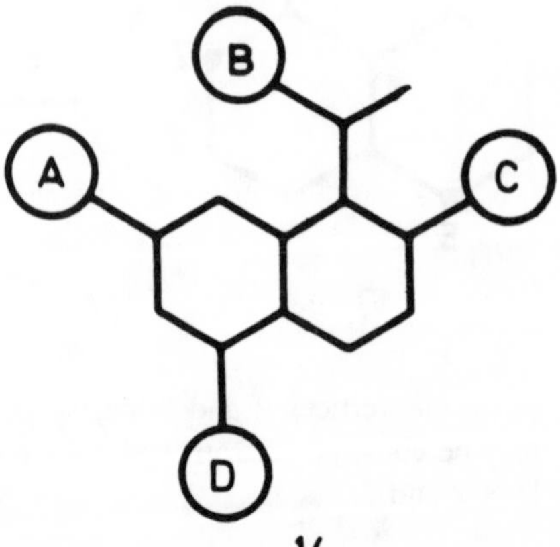

14

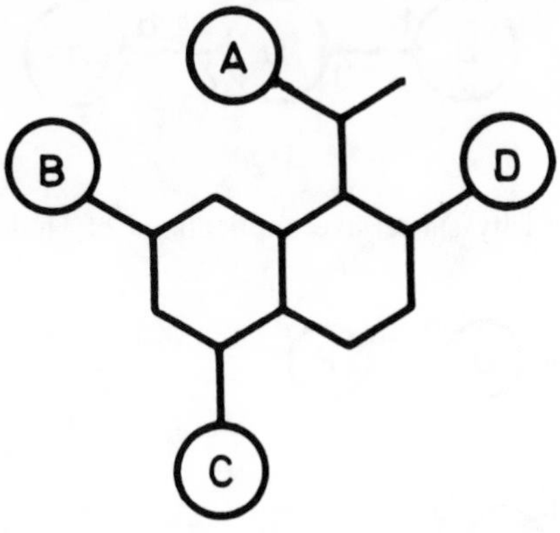

15

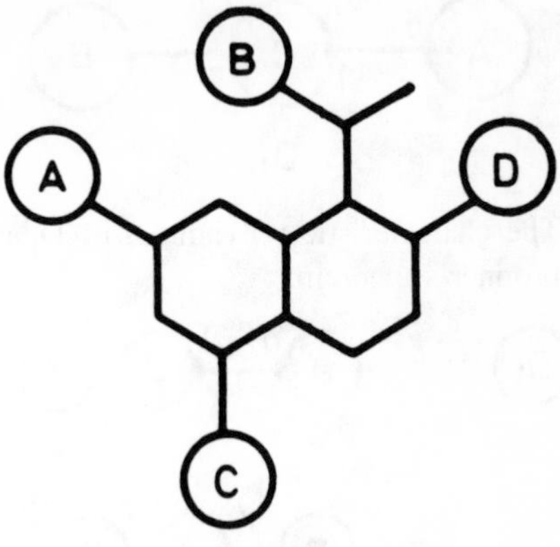

16

Frequently, one (or both) of the isospectral graphs derived from a common frame contain its own isospectral points at which additional fragments may be substituted in a reciprocal fashion to generate a new common-frame isospectral pair. The styrene derivative **2,** for example, contains a pair of isospectral points, denoted here by 2 and 8.

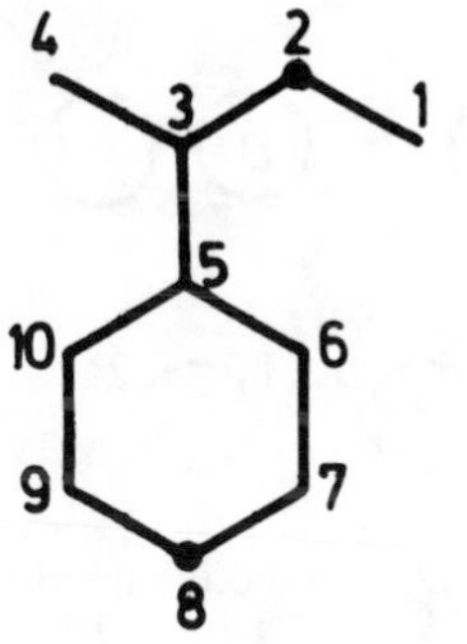

2

The vertices 2 and 8 are isospectral points by virtue of the fact that the subgraphs **2**-v_2 and **2**-v_8 are isospectral.

It should be noted at this point that, although 8 in **2** is the same vertex as 6 in the styrene frame, it is not generally true that isospectral points retain their unique properties in common-frame derivatives; here, for example, removal of v_3 from **2** does *not* produce a graph which is isospectral with or identical to the graphs **2**-v_2 and **2**-v_8.

The isospectrality of common-frame graphs can be accounted for by an application[35] of Heilbronner's decomposition theorem,[27] as stated in Equation (1). Consider two graphs LD_1 and D_2, derived by the reciprocal attachment of two fragments (A and B) to the isospectral points (1 and 2) of a common frame D, as illustrated below.

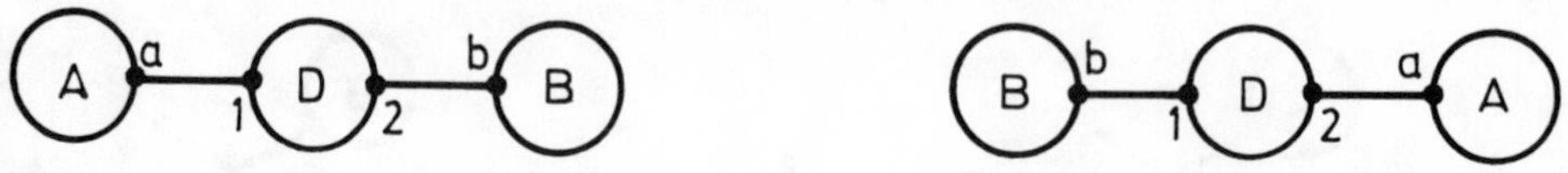

The characteristic polynomial of D_1 may be obtained by successive applications of Heilbronner's theorem:[27]

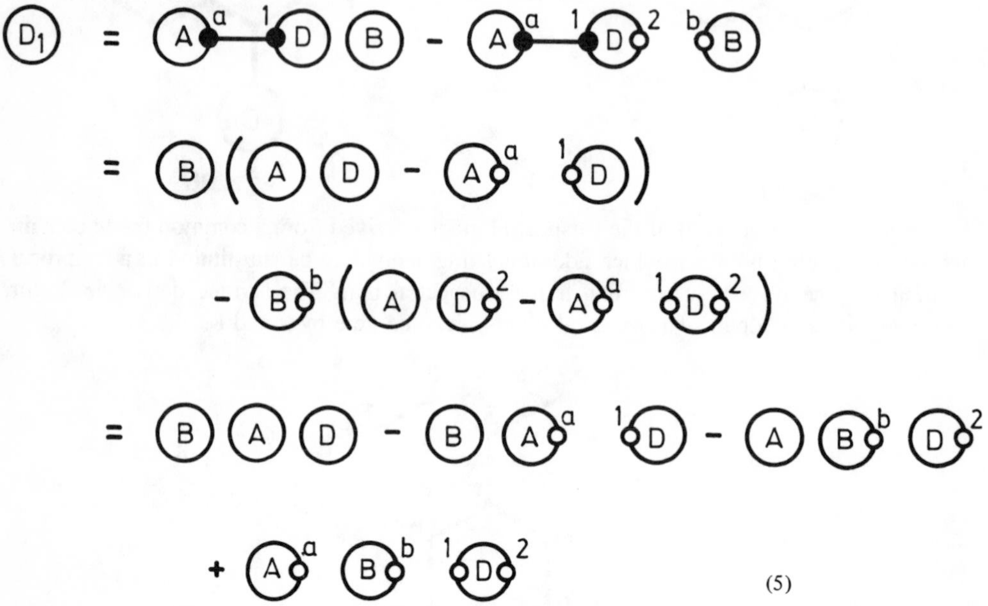

(5)

The characteristic polynomial of D_2 may be obtained in the same manner:

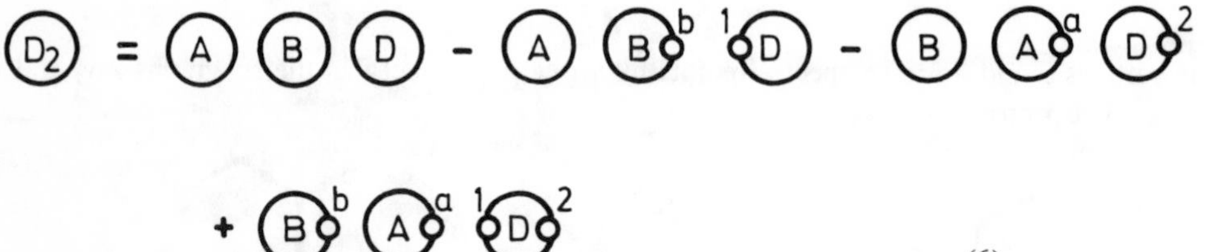

(6)

Since the graphs D-v_1 and D-v_2 are either isospectral or identical, the characteristic polynomials of the graphs D_1 and D_2 are identical term-by-term; therefore D_1 and D_2 are isospectral.

B. Isospectral Graphs Derived from an Isospectral Pair

Consider a pair of graphs G_1 and G_2 which are known to be isospectral. (These graphs may be derived from a common frame or they may be structurally unrelated.) Suppose that it is possible to identify a vertex a_1 in G_1 and a vertex a_2 in G_2 such that the deletion of these vertices and their adjoining edges produces two identical or isospectral graphs. In such a case, a_1 and a_2 constitute a pair of *substitution partners* in the sense that a new pair of isospectral graphs may be generated by the substitution of an arbitrary graphical fragment R into G_1 at vertex a_1, and into G_2 at vertex a_2.

The isospectral graphs **17** and **18** have been studied extensively by Randić and coworkers,[15]

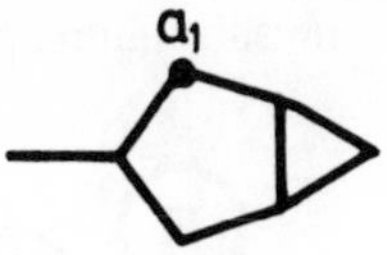

17

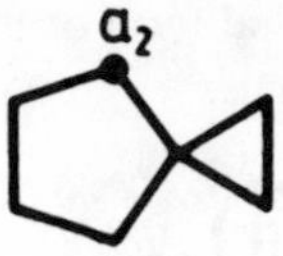

18

The vertices labeled a_1 and a_2 serve as substitution partners for these graphs, since the graphs **17**-a_1 and **18**-a_2 are identical.

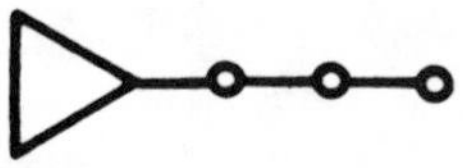

17-a_1 ≡ 18-a_2

From the twofold symmetry of **17** and **18,** it is evident that these particular graphs contain a second pair of substitution partners equivalent to a_1 and a_2; these may be denoted by a_3 (in **17**) and a_4 (in **18**).

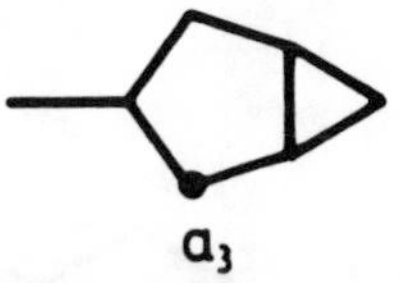

17

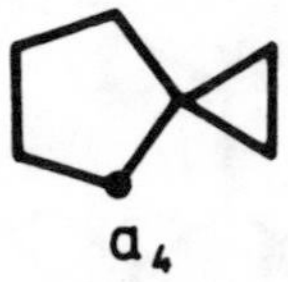

18

It is therefore possible to attach two arbitrary fragments R and S to **17** at vertices a_1 and a_3 and to **18** at vertices a_2 and a_4 to obtain two families of graphs which are isospectral for any given fragments R and S.

Further inspection of **17** and **18** reveals yet another pair of substitution partners, for deletion of the vertices labeled b_1 and b_2 produces the graph shown below.

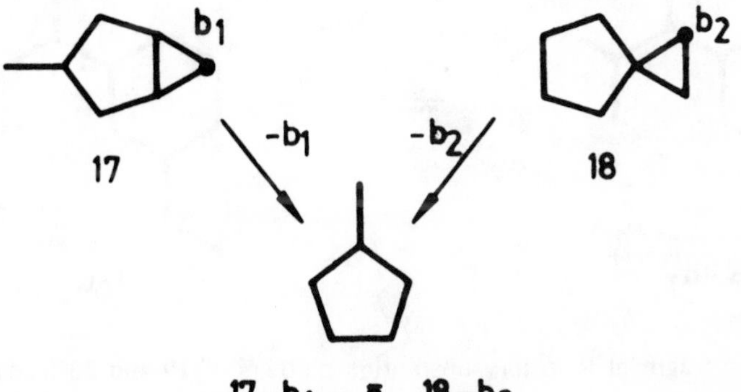

17-b_1 ≡ 18-b_2

The most general form of the isospectral families derivable by substitution from the graphs **17** and **18** is thus:

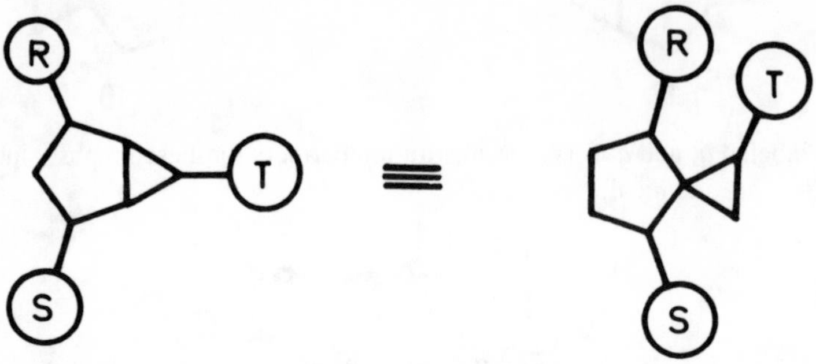

The common-frame graphs **19** and **20** contain vertices which are substitution partners by virtue of the fact that the removal of these vertices produces two isospectral graphs, rather than two identical graphs.

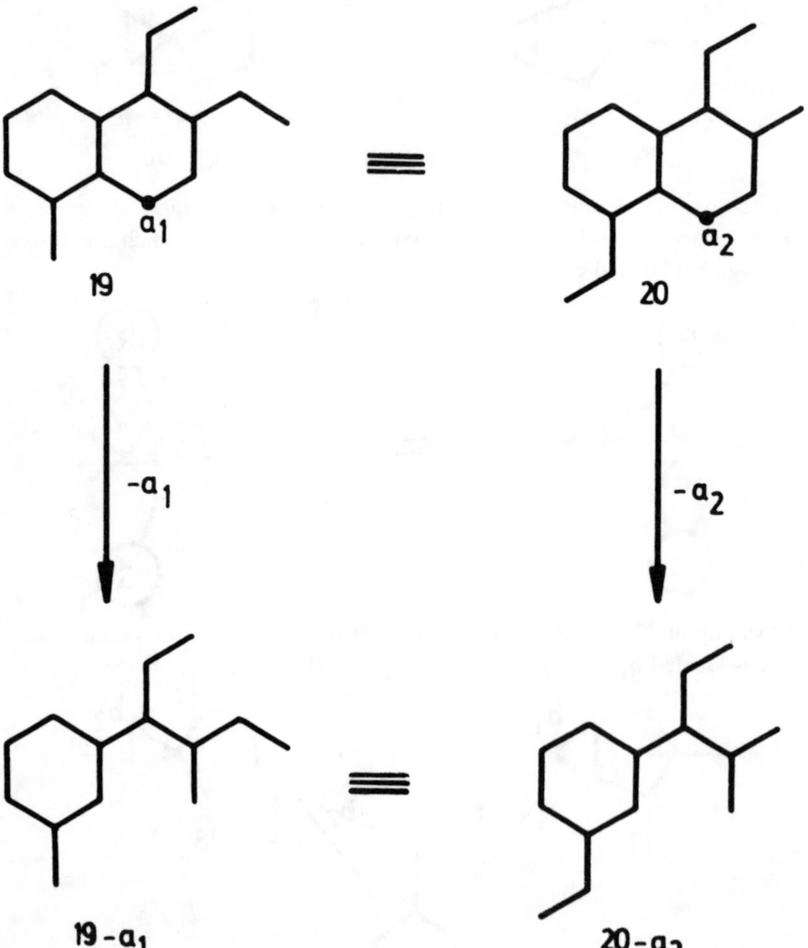

Attachment of a fragment R to the substitution partners of **19** and **20** leads to isospectral graphs of the form:

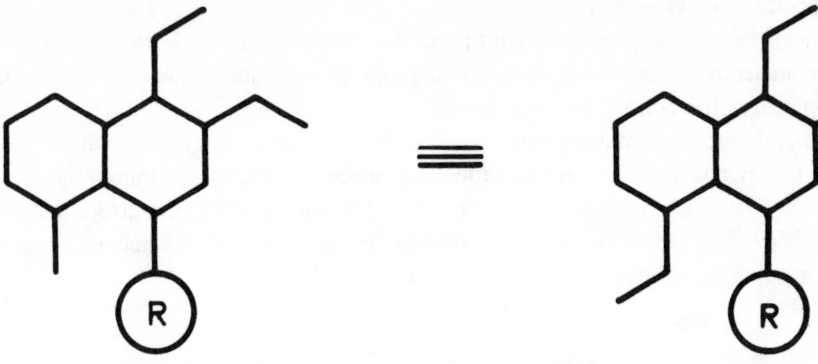

The generic styrene derivatives **8** and **9** have been studied at length both by Herndon and Ellzey[14] and by Živković et al.[12] These graphs contain two pairs of symmetry-equivalent substitution partners which may be used to construct isospectral families of the form:

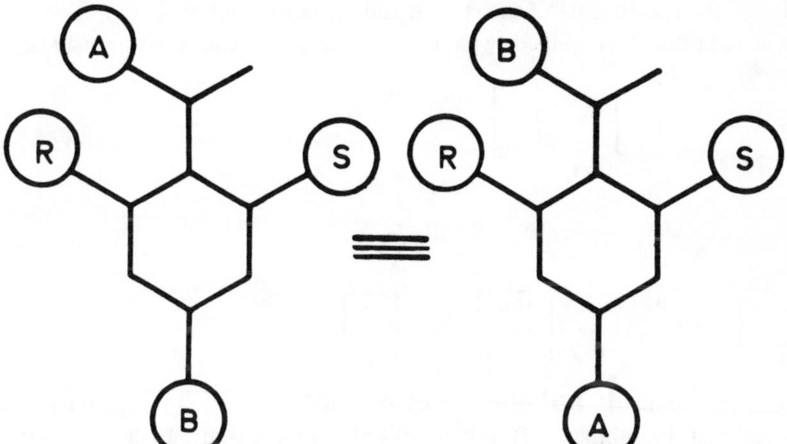

(Recall that A and B represent arbitrary graphical fragments in **8** and **9**.) It should be noted that vertices comprising substitution partners are called *unrestricted substitution points* by Herndon and Ellzey,[14] whereas they are called *active sites* by Živković et al.[12] in the case of **17** and **15** and *inactive sites* in the case of **8** and **9.**[12]

The isospectrality of a pair of graphs derived by substitution from isospectral parents can be accounted for by an application of the simplified form of Heilbronner's decomposition theorem[27] given in Equations (1) and (2).

Let G_2 (a_1) ——— $R(r)$ and $G_2(a_2)$ ——— $R(r)$ denote the graphs constructed by substitution of the fragment R at vertices a_1 and a_2 in the isospectral graphs G_1 and G_2. Heilbronner's formula gives as the characteristic polynomial of the substituted graphs:

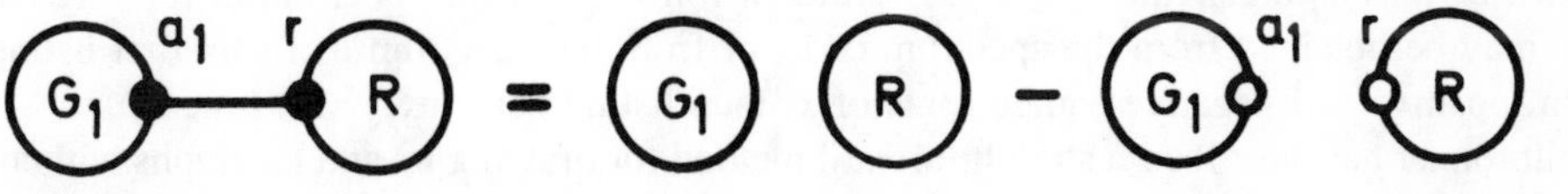

(7)

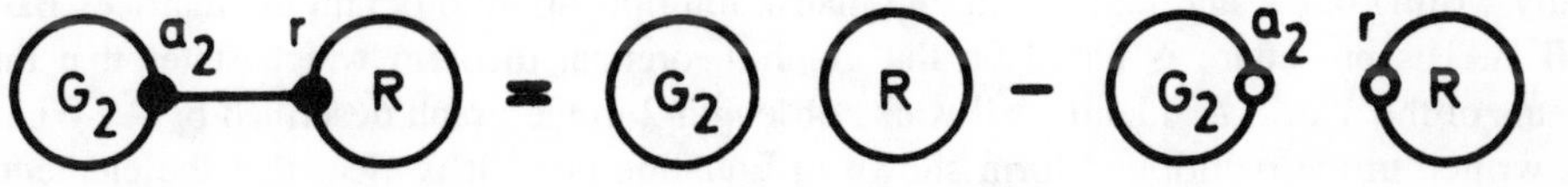

(8)

Since the parent graphs G_1 and G_2 are known to be isospectral, and since G_1-a_1 and G_2-a_2 are either isospectral or identical, the characteristic polynomials of the substituted graphs are identical, and thus the substituted graphs are isospectral.

C. The Procedure of Heilbronner

Heilbronner[22] has developed an elegant procedure which is useful both for accounting for observed instances of isospectrality between and among bipartite graphs and for constructing isospectral partners for certain bipartite graphs.

Suppose that the vertices of a bipartite graph are numbered in such a way that the "starred" vertices are numbered 1,2, . . . n_* and the "unstarred" vertices are numbered $(n_* + 1)$, $(n_* + 2)$, . . . , $(n_* + n_0)$, where n_* and n_0 represent the number of starred and unstarred vertices and, for convenience, n_* is less than or equal to n_0. The adjacency matrix of a bipartite graph numbered in this manner has the form,[36,37]

$$\mathbf{A} = \begin{bmatrix} \mathbf{0} & \mathbf{B} \\ \mathbf{B}^T & \mathbf{0} \end{bmatrix} \tag{9}$$

since vertices in the starred set are adjacent only to vertices in the unstarred set, and vice versa. The eigenvector of **A** also assumes a partitioned form $(u\ v)^T$, where u and v represent, respectively, the coefficients of the starred and unstarred vertices.

Suppose next that **A** is applied twice to its eigenvector. The equation may be written as,

$$\mathbf{A}^2 \begin{bmatrix} u \\ v \end{bmatrix} = \epsilon^2 \begin{bmatrix} u \\ v \end{bmatrix} \tag{10}$$

or as

$$\begin{bmatrix} \mathbf{BB}^T & 0 \\ 0 & \mathbf{B}^T\mathbf{B} \end{bmatrix} \begin{bmatrix} u \\ v \end{bmatrix} = \epsilon^2 \begin{bmatrix} u \\ v \end{bmatrix} \tag{11}$$

Notice that the submatrix $\mathbf{BB}^T$ of the block-diagonal matrix $\mathbf{A}^2$ belongs only to the starred vertices and that the submatrix $\mathbf{B}^T\mathbf{B}$ belongs only to the unstarred vertices. Performance of the matrix multiplication produces two equations, the first of which refers to the starred vertices alone and the second of which refers to the unstarred vertices alone.

$$\mathbf{BB}^T u = \epsilon^2 u \tag{12}$$

$$\mathbf{B}^T\mathbf{B} v = \epsilon^2 v \tag{13}$$

Consider now a graph L_* constructed in such a way that its adjacency matrix is given by $\mathbf{BB}^T$. Such a graph must contain n_* vertices and must have n_* positive eigenvalues which are related to the eigenvalues of the original graph by $x = \epsilon^2$. Consider a graph L_0 constructed in a similar fashion, so that its adjacency matrix is given by B^TB. L_0 must contain n_0 vertices and its spectrum must be identical to that of L_* except that the spectrum of the former contains zero as an eigenvalue $(n_0 - n_*)$ times. It follows that the spectrum of the original graph may be obtained from the spectrum of L_* or from the spectrum of L_0 by first listing the $2n_*$ positive and negative square roots of x, then listing zero $(n_0 - n_*)$ times.[38]

Heilbronner has described a straightforward method for drawing L_* and L_0 graphs without actually writing out A and performing the matrix multiplication to obtain the matrices $\mathbf{BB}^T$ and $\mathbf{B}^T\mathbf{B}$. His procedure is based on the graph-theoretical theorem which states that the elements of the matrix $\mathbf{A}^2$ identify all walks of length 2 in the graph described by A. When $\mathbf{A}^2$ is written in the partitioned form shown in Equation (10), it is clear that the elements $\mathbf{BB}^T$ describe the walks of length 2 from each starred vertex to each of the other starred vertices and that the elements of $\mathbf{B}^T\mathbf{B}$ list the walks of length 2 from each unstarred vertex to each of the other unstarred vertices. Thus, the i, jth element of $\mathbf{BB}^T$ is equal to the number of different ways in which one may reach the jth starred vertex from the ith starred vertex in two *steps* (this number equals the number of unstarred vertices that are connected with both *i* and *j*). The *i*, *i*th element of $\mathbf{BB}^T$ corresponds to the degree of the *i*th starred vertex, since one may describe a walk of length 2 by starting at vertex *i*, *stepping out* to each neighbor of *i*, then *stepping back* to *i*. The elements of $\mathbf{B}^T\mathbf{B}$ are defined in an analogous manner.

These considerations give rise to a three-step procedure for constructing L_* and/or L_0 for any bipartite graph G which contains no 4-membered cycles:

1. The vertices belonging to the starred and unstarred sets are rewritten separately.
2. Those vertices that are connected with the same vertex *v* of the other color are linked by an edge whose weight equals the number of such vertices *v*.
3. A weight equal to its degree in G is assigned to each vertex in the new graphs.

Heilbronner's construction process is applied to the styrene graph in the illustration below.

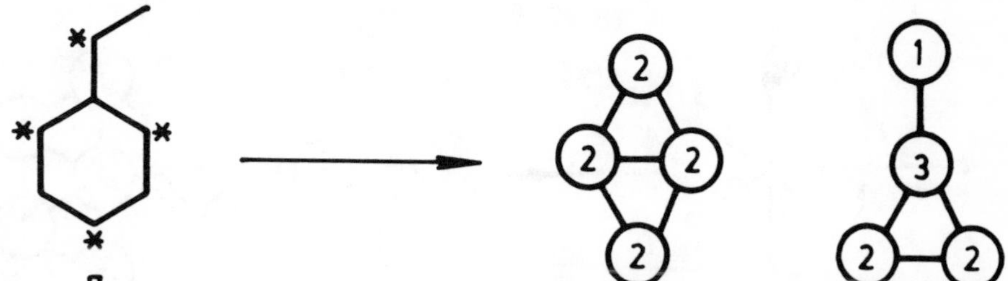

7 $L_*(7)$ $L_0(7)$

By constructing L_* and L_0, one may rapidly verify the existence of an isospectral relationship between two bipartite graphs having the same total number of vertices. If at least one of the two graphs L_*, L_0 of a graph G is identical to or isospectral with at least one of the graphs L'_*, L'_0 of G′, then G and G′ are isospectral. A comparison of the L_* and L_0 graphs of the styrene derivatives **1** and **2**, for example, shows immediately that **1** and **2** are isospectral.

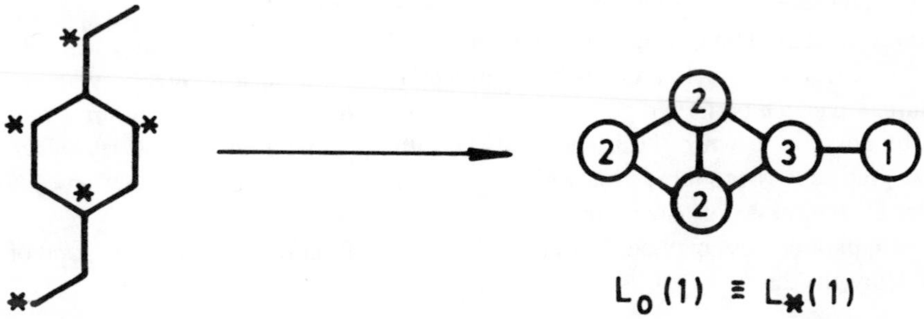

$L_0(1) \equiv L_*(1)$

1

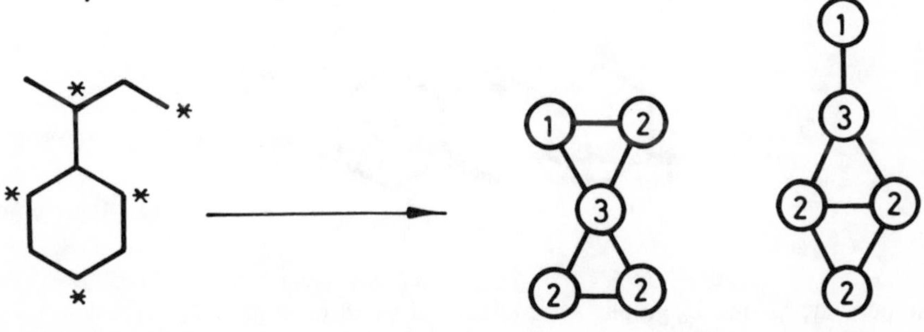

2 $L_0(2)$ $L_*(2)$

It is slightly more difficult to confirm the isospectrality of the graphs **21** and **22**, since neither of the graphs L_* (**21**), L_0 (**21**) is identical to L_* (**22**) or L_0 (**22**).

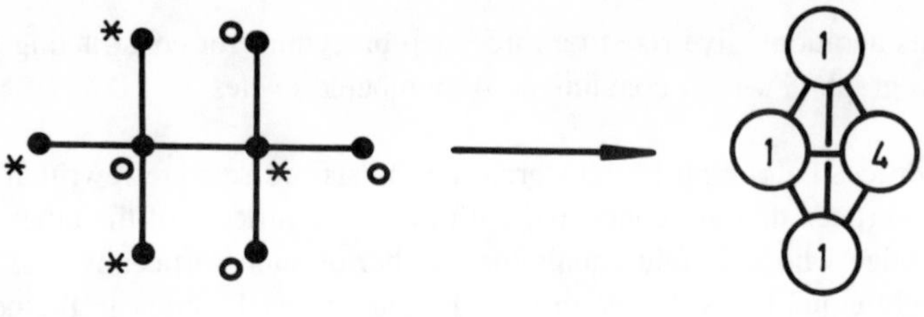

21 $L_x(21) = L_0(21)$

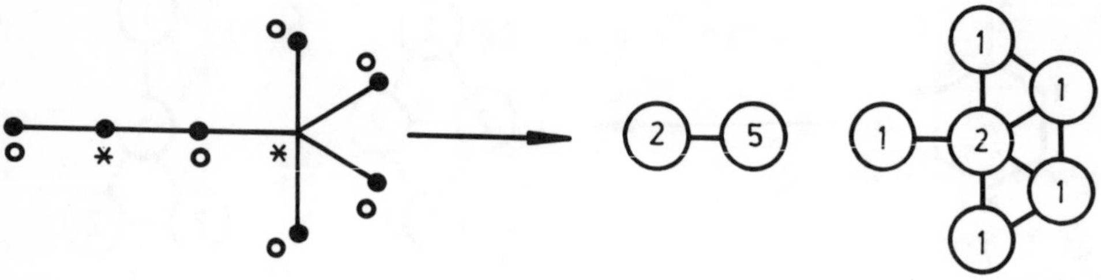

22 $L_x(22)$ $L_0(22)$

In this case, the spectra of the constructed graphs must actually be computed; nonetheless, the calculation involved is much simpler that required to obtain the spectrum of **21** and **22**, since the determinants which must be expanded are of order 2 × 2 and 4 × 4, rather than 8 × 8 (recall that it is only necessary to compute the $2n_*$ positive and negative square roots of the n_* eigenvalues of L_*, then to list zero as an eigenvalue $n_0 - n_*$ times).

Constructed starred and unstarred graphs (as L_* and L_0 may be called) may also be used, in many cases, to construct an isospectral partner for a given bipartite graph by a *wrapping* procedure. Given a bipartite graph G containing n_* starred vertices and n_0 unstarred vertices, it may be possible to *wrap* the L_* graph of G with n_0 vertices in such a way that a new bipartite graph G′ is produced, for which $L'_* = L_*$. In cases for which this procedure is possible, G′ is necessarily isospectral with G.

The wrapping process may be illustrated by its application to the molecular graph of 1,1-diphenylethylene **23**.

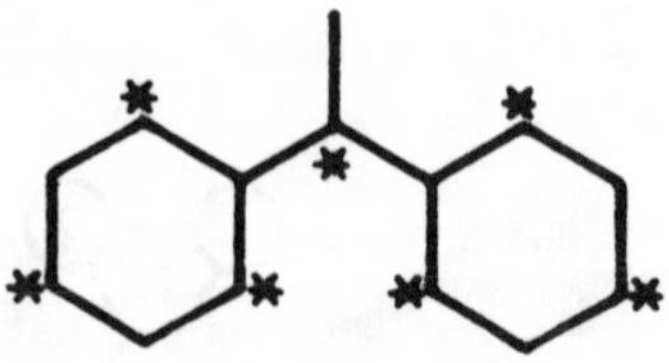

23

As a first step, the L_* and L_0 graphs are constructed by inspection of **23**.

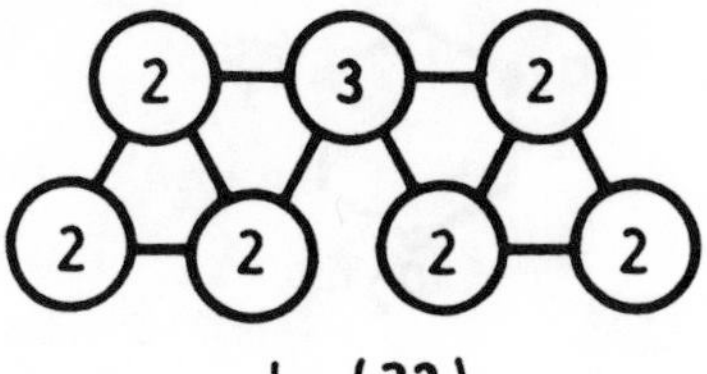

$L_*(23)$

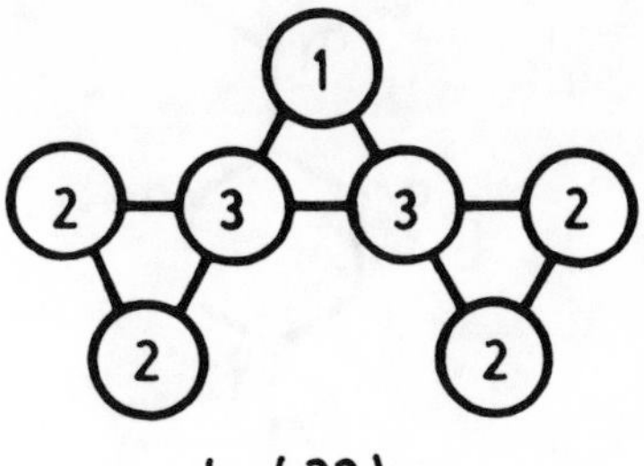

$L_0(23)$

Next, seven unstarred vertices are placed among the vertices of $L_*(\mathbf{23})$ in such a way that their edge pattern differs from that of $L_0(\mathbf{23})$.

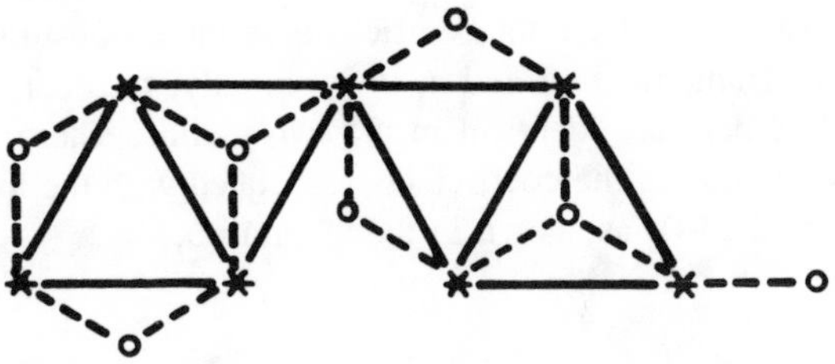

$L_*(23)$, wrapped

A new graph **24** is now constructed by joining the starred and unstarred vertices in the wrapped form of $L_*(\mathbf{23})$.

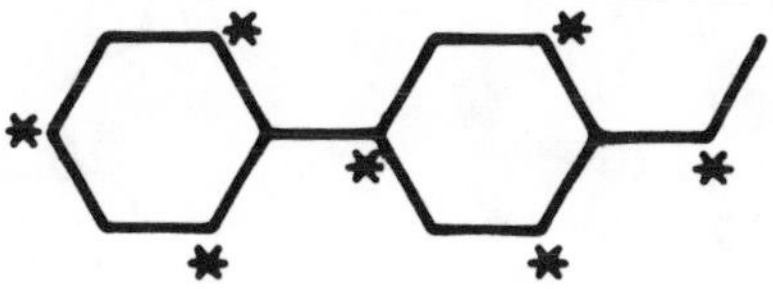

24

Since the L_* graph of **24** is identical to the L_* graph of **23, 24** must be an isospectral partner of **23**. An examination of compilations of π-electron eigenvalues reveals that the common-frame molecular graphs representing 1,1-diphenylethylene and 4-vinylbiphenyl are, in fact, isospectral.[32-34]

III. COEFFICIENT REGULARITIES IN ISOSPECTRAL GRAPHS

Investigations of isospectral graphs have revealed many interesting regularities among the eigenvectors of the adjacency matrices of such graphs. Since the topological eigenvectors are identical to the molecular orbitals obtained by a simple Hückel MO treatment,[38,39] their observed regularities may be related to molecular properties, such as bond order, charge density, and free valence, which can be calculated from the Hückel MOs.

The eigenvectors of the 1,4-divinylbenzene (**1**) and 2-phenylbutadiene (**2**) graphs have been studied in detail by Živković et al.[12]

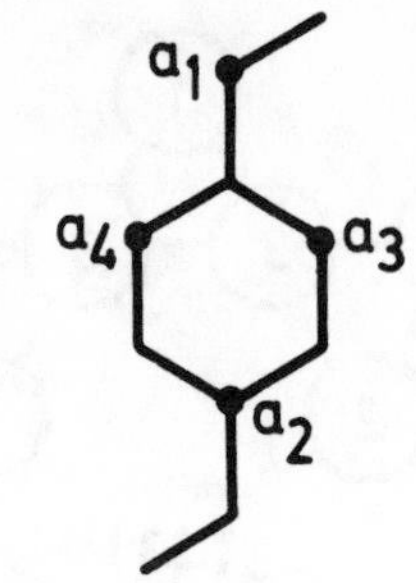

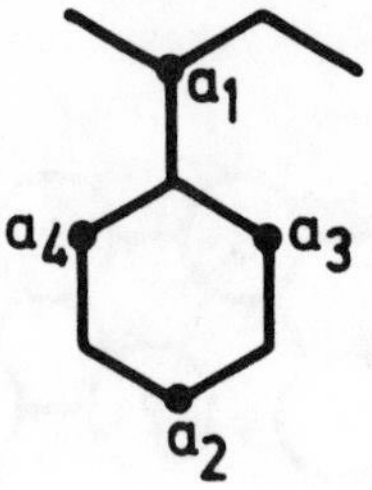

1 2

These authors have pointed out that the coefficients of the vertices (a_1, a_2, a_3, a_4) which serve as substitution partners for these common frame isospectral graphs are identical (to within a sign) for a given eigenvalue. Furthermore, they note that the coefficients of the isospectral point of the common frame to which the ethylene fragment has been attached (a_2 in **1** and a_1 in **2**) are equal, as are the coefficients of the unsubstituted isospectral points of the common styrene frame (a_1 in **1** and a_2 in **2**). Finally, the coefficients of the vertices within the ethylene fragment are identical in the two graphs. These relationships may be seen more clearly by examining the coefficients associated with the lowest common eigenvalue of **1** and **2** ($x_1 = 2.214$), as shown in the diagram below.

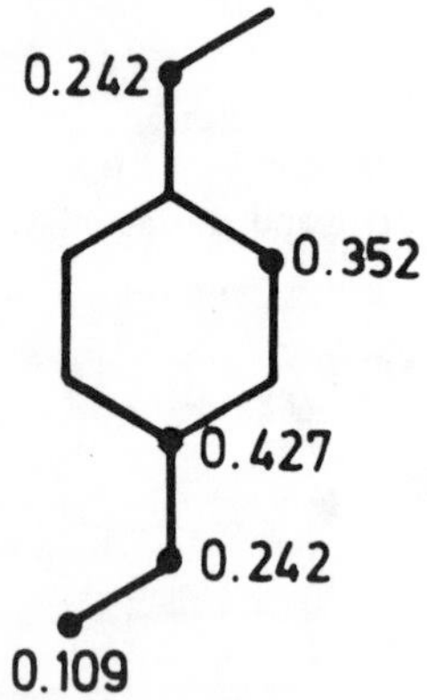

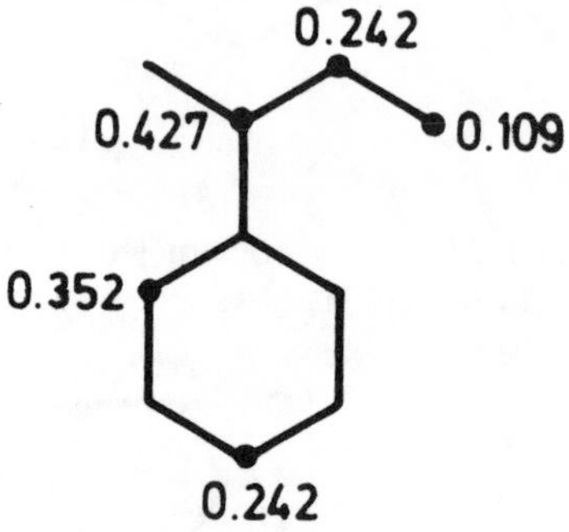

1 2

The relationships among the coefficients of the vertices in a generalized pair of isospectral styrene derivatives may be diagrammed, with arrows linking vertices whose coefficients are equal.

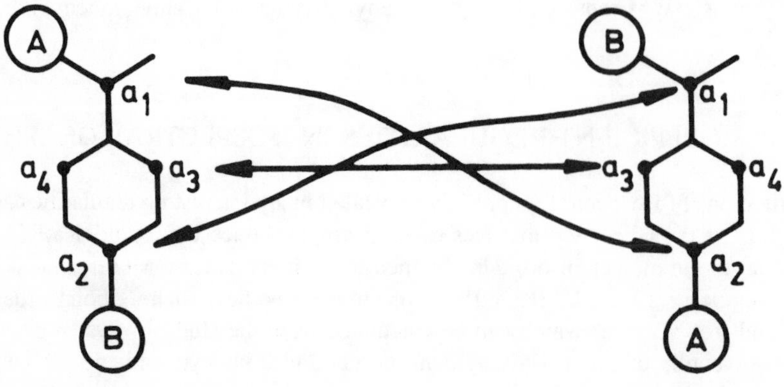

8 9

The alternation properties of the styrene frame (indicated in the diagram above by enlarged vertices) provide an explanation for the observed coefficient regularities. By applying Heilbronner's construction procedure to **8** and **9**, it is immediately evident that these graphs have the same L_* graph.

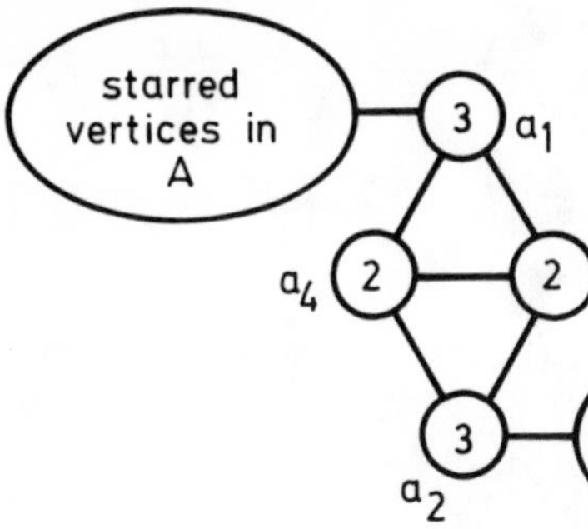

L_* (8)

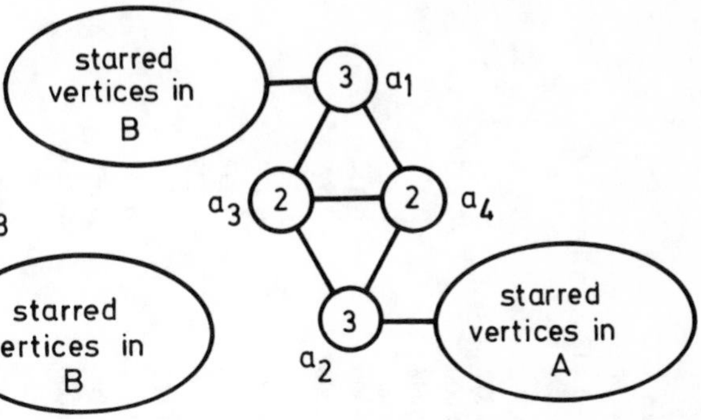

L_* (9)

Clearly, the vertices a_3 and a_4 in L_*(**8**) are equivalent to vertices a_3 and a_4 in L_*(**9**), just as a_1 and a_2 in L_*(**8**) are respectively equivalent to a_2 and a_1 in L_*(**9**); thus, these pairs of equivalent vertices necessarily have the same coefficient in *u*, the eigenvector of the adjacency matrix of the L_* graph. Coefficient regularities in bipartite isospectral graphs sharing a frame other than styrene may be explained by a similar analysis.

Coefficient regularities among structurally unrelated isospectral graphs have been also detected. Several regularities appear among the coefficients of the vertices of the graphs **17** and **18** (see Table 1).

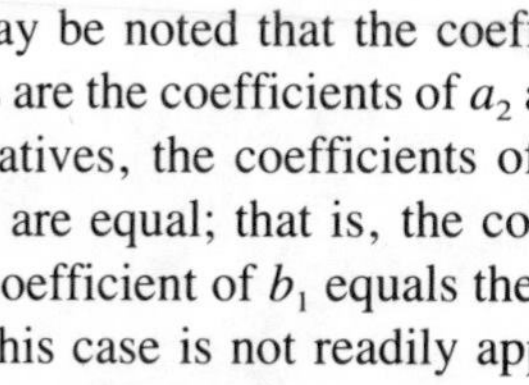

17

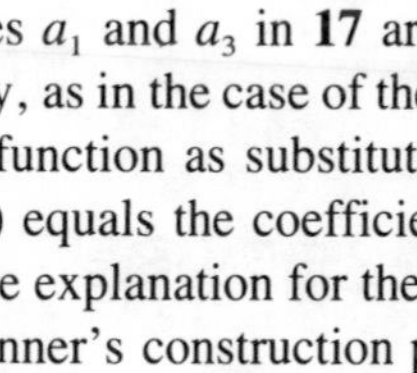

18

First, it may be noted that the coefficients of the vertices a_1 and a_3 in **17** are related by symmetry, as are the coefficients of a_2 and a_4 in **18**. Secondly, as in the case of the isospectral styrene derivatives, the coefficients of the vertices which function as substitution partners in **17** and **18** are equal; that is, the coefficient of a_1 (or a_3) equals the coefficient of a_2 (or a_4), and the coefficient of b_1 equals the coefficient of b_2. The explanation for the regularities observed in this case is not readily apparent, since Heilbronner's construction procedure is not applicable to **17** and **18**.

Another interesting feature of the coefficients of the two graphs may be noted. It has been observed by D'Amato[40,41] that several *additivity* relationships exist among the coefficients of vertices in **17** and **18** and that such relationships may be predicted from certain features of the graphs **17**-a_1 = **18**-a_2 and **17**-b_1 = **18**-b_2. These relationships and features wil now be described in some detail.

17 – a_1

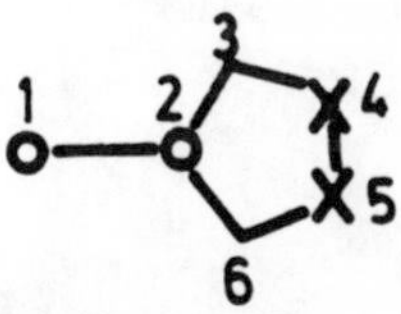

17 – b_1

Table 1
COEFFICIENTS OF THE VERTICES IN GRAPHS 17 AND 18

Vertex in 17	$x_1 = 2.473$	$x_2 = 1.463$	$x_3 = 0.618$	$x_4 = 0.000$	$x_5 = -1.000$	$x_6 = -1.618$	$x_7 = -1.935$
a_1	0.331	0.247	0.602	0.378	0.000	−0.372	0.432
a_3	0.331	0.247	−0.602	0.378	0.000	0.372	0.432
b_1	0.404	−0.373	0.000	−0.378	0.707	0.000	0.235
c_1	0.499	−0.273	0.372	0.000	−0.354	0.602	−0.227
c_2	0.499	−0.273	−0.372	0.000	−0.354	−0.602	−0.227
c_3	0.130	0.433	0.000	−0.756	−0.354	0.000	0.314
c_4	0.320	0.634	0.000	−0.000	0.354	0.000	−0.609
Vertex in 18							
a_2	0.331	0.247	0.602	−0.378	0.000	−0.372	−0.432
a_4	0.331	0.247	−0.602	−0.378	0.000	0.372	−0.432
b_2	0.404	−0.373	0.000	0.378	0.707	0.000	−0.235
d_1	0.225	0.534	0.372	0.378	0.000	0.602	0.147
d_2	0.225	0.534	−0.372	0.378	0.000	−0.602	0.147
d_3	0.594	−0.173	0.000	−0.378	0.000	0.000	0.688
d_4	0.404	−0.373	0.000	0.378	−0.707	0.000	−0.235

Table 2
COEFFICIENTS OF THE VERTICES IN GRAPH 17 $-a_1$

Vertex	$x_1 = 2.228$	$x_2 = 1.360$	$x_3 = 0.186$	$x_4 = -1.000$	$x_5 = -1.000$	$x_6 = -1.775$
1	0.090	0.485	−0.632	0.535	0.000	0.267
2	0.201	0.660	−0.118	−0.535	0.000	−0.474
3	0.357	0.413	0.610	0.000	0.000	0.574
4	0.595	0.099	0.231	0.535	0.000	−0.545
5	0.485	−0.273	−0.284	−0.267	0.707	0.196
6	0.485	−0.273	−0.284	−0.267	−0.707	0.196

Table 3
COEFFICIENTS OF THE VERTICES IN GRAPH 17$-b_1$

Vertex	$x_1 = 2.115$	$x_2 = 1.000$	$x_3 = 0.618$	$x_4 = -0.254$	$x_5 = -1.618$	$x_6 = -1.861$
1	0.247	0.500	0.000	0.749	0.000	0.357
2	0.523	0.500	0.000	−0.190	0.000	−0.664
3	0.429	0.000	0.602	−0.351	−0.372	0.439
4	0.385	−0.500	0.372	0.280	0.602	−0.153
5	0.385	−0.500	−0.372	0.280	−0.602	−0.153
6	0.429	0.000	−0.602	−0.351	0.372	0.439

Additivity relationships among coefficients in a single graph. In the graph **17**-a_1, it is observed that the sum of the coefficients of vertices 2 and 6 is equal to the sum of the coefficients of vertices 1 and 4. (The coefficients of all vertices in this graph are listed in Table 2.)

The vertices whose coefficients are to be added together have been labeled with *x* s and *o* s for ready identification, the sum of the *x*-marked vertices equals the sum of the *o*-marked vertices.

In the graph 17-b_1, a similar relationship is found to exist between the sum of the coefficients of vertices 1 and 2 and the sum of the coefficients of vertices 4 and 5 (see Table 3.)

The additivity relationships in both cases may be explained by an analysis of the matrix equation

$$\mathbf{A\,C} = \mathbf{C\,X} \tag{14}$$

where **A** represents the adjacency matrix of a given graph, **C** represents the matrix of its eigenvectors, and **X** represents the diagonal matrix of its eigenvalues. Examining a particular element of the matrix product on each side of equation (14) one obtains:

$$(\mathbf{A\,C})_{ij} = \sum_m (\mathbf{A})_{im}\,(\mathbf{C})_{mj} \tag{15}$$

$$(\mathbf{C\,X})_{ij} = \sum_m (\mathbf{C})_{im}\,(\mathbf{X})_{mj} = (\mathbf{C})_{ij}\,(\mathbf{X})_{jj} \tag{16}$$

Equations (15) and (16) may be combined to give

$$\sum_m (\mathbf{A})_{im}\,(\mathbf{C})_{mj} = (\mathbf{C})_{ij}\,(\mathbf{X})_{jj} \tag{17}$$

For a given value of *j*, Equation (17) reduces to

$$\sum_m (\mathbf{A})_{im}\,c_m = c_i\,x \tag{18}$$

where c_i is the coefficient of vertex i in the jth eigenvector c_j. A particular term in the summation on the left-hand side of Equation (18) is zero if no edge exists between vertices i and m, whereas it has the value $1 \cdot c_m$ if an edge does exist between vertices i and m. For every vertex in the graph of interest, then, there is an equation that relates the product of x and the coefficient of that vertex to the sum of the coefficients of all the vertices adjacent to that vertex.

When the sum rule defined by Equation (18) is applied to the graph **17**-a_1, the following set of equations is obtained:

$$c_2 = c_1 x \quad (19)$$

$$c_1 + c_3 = c_2 x \quad (20)$$

$$c_2 + c_4 = c_3 x \quad (21)$$

$$c_3 + c_5 + c_6 = c_4 x \quad (22)$$

$$c_4 + c_6 = c_5 x \quad (23)$$

$$c_4 + c_5 = c_6 x \quad (24)$$

A comparison of the sum of equations (19) and (22) with the sum of equations (20) and 24) — or, equivalently, (20) and (23) — leads immediately to the conclusion that $c_1 + c_4 = c_2 + c_6$ (or $c_2 + c_5$).

$$(c_2 + c_6) + (c_3 + c_5) = x\,(c_1 + c_4) \quad (19) + (22)$$

$$(c_1 + c_4) + (c_3 + c_5) = x\,(c_2 + c_6) \quad (20) + (24)$$

The additivity relationships observed in the graph **17**-b_1 may also be explained by, first, applying the sum rule to each vertex to generate six equations:

$$d_2 = d_1 x \quad (25)$$

$$d_1 + d_3 + d_6 = d_2 x \quad (26)$$

$$d_2 + d_4 = d_3 x \quad (27)$$

$$d_3 + d_5 = d_4 x \quad (28)$$

$$d_4 + d_6 = d_5 x \quad (29)$$

$$d_2 + d_5 = d_6 x \quad (30)$$

then comparing the sum of Equations (25) and (26) with the sum of Equations (28) and (29).

$$(d_1 + d_2) + (d_3 + d_6) = x\,(d_1 + d_2) \quad (25) + (26)$$

$$(d_4 + d_5) + (d_3 + d_6) = x\,(d_4 + d_5) \quad (28) + (29)$$

Clearly, $d_1 + d_2 = d_4 + d_5$.

Predicting additivity relationships within a graph. The question arises as to how the additivity relationships observed in the graphs **17**-a_1 and **17**-b_1 might be predicted on the basis of structural features alone. Several guidelines for making such a prediction may be stated.

Examine again the graph **17**-b_1. Notice that the x-marked vertices are connected to vertices 3 and 6, as are the o-marked vertices. Furthermore, the sum of the degrees of the x-marked vertices (four) is equal to the sum of the degrees of the o-marked vertices. These two conditions guarantee that the sum-rule Equations (25) and (26) for vertices 1 and 2 together contain the same number of terms as the sum-rule Equations (28) and (29) for vertices 4 and 5 and that, moreover, both sets of equations contain the terms d_3 and d_6.

In the case of **17**-a_1, the o-marked vertices are connected to vertices 2,3,5, and 6, and the x-marked vertices are connected to vertices 1,3,4, and 5. Again, the sums of the degrees of the x- and o-marked vertices are equal. In this instance, however, the x- and o-vertices are not only connected to a common pair of vertices, as was the case in **17**-b_1, but also connected in a reciprocal fashion to one another. The sum-rule equations for vertices 1 and 4 and the equations for vertices 2 and 6 thus contain c_3 and c_5 as common terms, and they contain the terms $(c_1 + c_4)$ and $(c_2 + c_6)$ in a reciprocal relationship.

Unfortunately, the process of searching a graph for the existence of vertices fulfilling these conditions is very tedious. The guidelines stated here may be more useful in explaining observed additivity relationships among coefficients than in predicting new ones.

Coefficient regularities in **17** and ***18***. The addivity relationships in the graphs **17**-a_1 and **17**-b_1 may now be related to certain coefficient regularities in **17** and **18**.

Suppose that the x-marked vertices in **17**-a_1 are bridged through a single vertex to produce the graph **17** and that the o-marked vertices are bridged through a single vertex to produce the graph **18.**

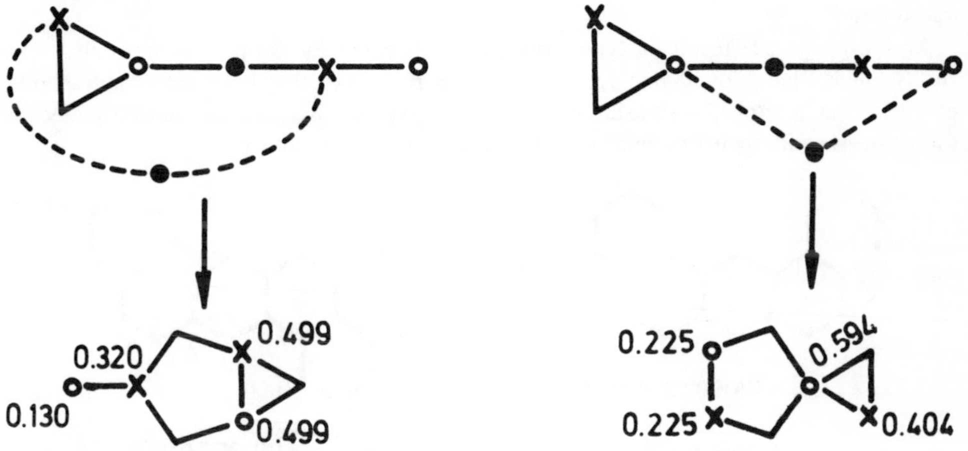

The value of the coefficients at the marked positions have been indicated. Notice that the sum of the coefficients of the x-marked vertices in **17** is identical to the sum of the coefficients of the o-marked vertices in **18** and that the reverse is also true.

The graphs **17** and **18** may be generated by bridging, through a single vertex, the x-marked and o-marked vertices (respectively) of **17**-b_1.

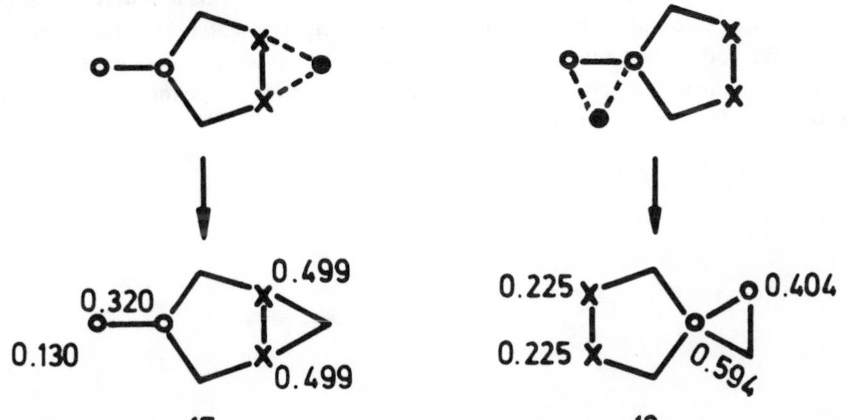

17 18

Again, the sum of the coefficients of the *x*-marked vertices in **17** equals the sum of the coefficients of the *o*-marked vertices in **18,** and vice versa.

IV. CHEMICAL BEHAVIOR OF ISOSPECTRAL MOLECULES

No survey of current research involving isospectral graph would be complete without mentioning chemical behavior of isospectral systems. One might expect to find that some properties of isospectral molecules are closely related. However, the experimental data on isospectral molecules are very limited indeed. So far, only photoelectron spectra of the isospectral pair: 1,4-divinylbenzene (**1**) and 2-phenylbutadiene (**2**) have been studied.[42] Heilbronner and Jones[42] have measured the ionization potentials of **1** and **2** and have reported that the measured values ''differ at least as much as those of any other *nonisospectral* pair having systems of comparable size''. This result is not unexpected because spectral properties of molecules are only partially topology-dependent.[43,44] (Spectral properties referred to here concern the physical spectra of molecules). Unfortunately, no other experimental evidence is available since instances of isospectrality among molecular graphs are fairly rare. Perhaps some relationships among the physical and chemical properties of isospectral molecules will be detected in the future, since such a possibility is now predicted, and could be, hopefully, traced down.

An intriguing side result has come out from the search for isospectral molecules. Clar and Schmidt[45] have found polycyclic conjugated molecules that have remarkable similar photoelectron spectra. As an example, consider a pair of polycyclic conjugated molecules shown below with their first and second ionization potentials (in eV).

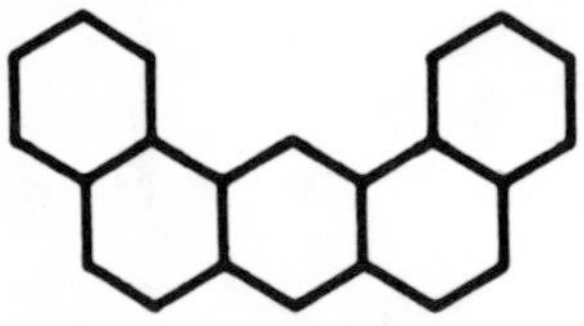

1. 2,7. 8 - dibenzanthracene

7.40/7.79

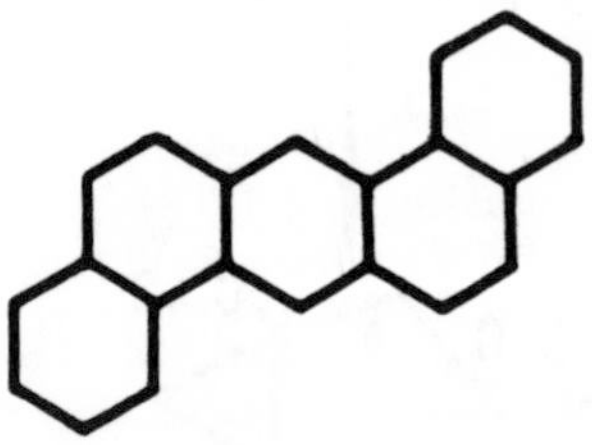

1.2,5.6 - dibenzanthracene

7.38/7.80

These molecules have different Hückel spectra. In order to avoid confusion with the term isospectral, Clar and Schmidt[45] refer to these molecules as isotopic.

In addition, the fact that the characteristic polynomial does not uniquely represent the topology of a molecule ended attempts to produce a sorting device, based on the characteristic polynomial for coding and retrieving chemical structures in computer-oriented systems.[46] However, the failure of this approach triggered attempts to find a graph theoretical polynomial or topological index which can distinguish isospectral molecules.[47-49]

REFERENCES

1. **Harary, F.,** *SIAM (Soc. Ind. Appl. Math.) Rev.,* 4, 202, 1962.
2. **Spialter, L.,** *J. Am. Chem. Soc.,* 85, 2012, 1963.
3. **Spialter, L.,** *J. Chem. Doc.,* 4, 261, 1964.
4. **Kudo, Y., Yamasaki, T., and Sasaki, S.-I.,** *J. Chem. Doc.,* 13, 225, 1973.
5. **Collatz, L. and Sinogowitz, U.,** *Abh. Math. Semin. Univ. Hamburg,* 21, 63, 1967.
6. **Balaban, A. T. and Harary, F.,** *J. Chem. Doc.,* 11, 258, 1971.
7. **Mizutani, V., Kawasaki, K., and Hosoya, H.,** *Nat. Sci. Rep. Ochanomizu Univ.,* 22, 39, 1971.
8. **Kawasaki, K., Mizutani, K., and Hosoya, H.,** *Nat. Sci. Rep. Ochanomizu Univ.,* 22, 181, 1971.
9. **Schwenk, A. J.,** in *New Directions in the Theory of Graphs,* Harary, F., Ed., Academic Press, New York, 1973, 275.
10. **Gutman, I. and Trinajstić, N.,** *Topics Curr. Chem.,* 42, 49, 1973.
11. **Harary, F.,** *Graph Theory,* Addison-Wesley, Reading, Mass., 1971, 158, second printing.
12. **Živković, T., Trinajstić, N., and Randić, M.,** *Mol. Phys.,* 30, 517, 1975.
13. **Herndon, W. C.,** *Tetrahedron Lett.,* 671, 1974.
14. **Herndon, W. C. and Ellzey, M. L., Jr.,** *Tetrahedron,* 31, 99, 1975.
15. **Randić, M., Trinajstić, N., and Živković, T.,** *J. Chem. Soc. Faraday Trans. 2,* 244, 1976.
16. **D'Amato, S. S., Gimarc, B. M., and Trinajstić, N.,** *Croat. Chem. Acta,* 54, 1, 1981.
17. **Baker, G. A., Jr.,** *J. Math. Phys.,* 7, 2238, 1966.
18. **Krishnamoorty, V. and Parthasarathy, K. R.,** *J. Comb. Theory,* (B)17, 39, 1974; (B)19, 204, 1975.
19. **Harary, F., King, C., Mowshowitz, A., and Read, R. C.,** *Bull. London Math. Soc.,* 3, 321, 1971.
20. **Mowshowitz, A.,** *J. Comb. Theory,* (B)12, 177, 1972.
21. **Schwenk, A. J., Herndon, W. C., and Ellzey, M. L., Jr.,** in Second International Conference on Combinatorial Mathematics, Gewirtz, A. and Quintas, L. V., Eds., *Ann. N.Y. Acad. Sci.,* 319, 490, 1979.
22. **Heilbronner, E.,** *Math. Chem. (Mülheim/Ruhr),* 5, 105, 1979.
23. **Sachs, H.,** *Publ. Math. (Debrecen),* 11, 119, 1964.
24. **Trinajstić, N.,** *Croat. Chem. Acta,* 49, 593, 1977.
25. **König, D.,** *Theorie der endlichen und unendlichen Graphen,* Akademische Verlagsgesellschaft, Leipzig, 1936.
26. **Coulson, C. A. and Rushbrooke, G. S.,** *Proc. Cambridge Phil. Soc.,* 36, 193, 1940.
27. **Heilbronner, E.,** *Helv. Chim. Acta,* 36, 170, 1953.
28. **Randić, M.,** *J. Comput. Chem.,* 1, 386, 1980.
29. **Sykes, M. F. and Fisher, M. E.,** *Adv. Phys.,* 9, 315 1960.
30. **Randić, M.,** *J. Chem. Phys.,* 60, 3920, 1974.
31. **Randić, M.,** *J. Chem. Phys.,* 62, 309, 1975.
32. **Streitwieser, A., Jr. and Brauman, J. I.,** *Supplemental Tables of Molecular Orbital Calculations,* Pergamon Press, Elmsford, N.Y., 1965.
33. **Coulson, C. A. and Streitwieser A., Jr.,** Dictionary of π-Electron Calculations, W. H. Freeman, San Francisco, 1965.
34. **Heilbronner, E. and Straub, P. A.,** *Hückel Molecular Orbitals,* Springer-Verlag, Berlin, 1966.
35. **Živković, T., Trinajstić, T., and Randić, M.,** *Croat. Chem. Acta,* 49, 89, 1977.
36. **Hall, G. G.,** *Proc. R. Soc. London, Ser. A,* 229, 251, 1955.
37. **Ham, N. S.,** *J. Phys. Chem.,* 29, 1228, 1958.
38. **Graovac, A., Gutman, I., and Trinajstić, N.,** *Topological Approach to the Chemistry of Conjugated Molecules, Lecture Notes in Chemistry,* Vol. 4, Springer-Verlag, Berlin, 1977.
39. **Trinajstić, N., in** *Semiempirical Methods of Electronic Structure Calculations. Part A: Techniques,* Vol. 7, Segal, G. A., Ed., Plenum Press, New York, 1977, 1.
40. **D'Amato, S. S.,** *Mol. Phys.,* 37, 1363, 1979.
41. **D'Amato, S. S.,** *Theor. Chim. Acta,* 53, 319, 1979.
42. **Heilbronner, E. and Jones, T. B.,** *J. Am. Chem. Soc.,* 100, 6506, 1978.
43. **Eilfeld, P. and Schmidt, W.,** *J. Electron Spectrosc. Relat. Phenom.,* 24, 101, 1981.
44. **Bonchev, D. and Mekenyan, O.,** Report at the Ninth National Conference on Molecular Spectroscopy, Albena (Bulgaria), September 29 to October 3, 1980.
45. **Clar, E. and Schmidt, W.,** *Tetrahedron,* 35, 2673, 1979.
46. **Herndon, W. C.,** *J. Chem. Doc.,* 14, 150, 1974.
47. **Bonchev, D. and Trinajstić, N.,** *Int. J. Quantum Chem.,* S 12, 293, 1978.
48. **Bonchev, D.,** Information Theoretic Indices for Characterization of Chemical Structures, Research Studies Press, Chichester, in press.
49. **Bonchev, D., Mekenyan, O., and Balaban, A. T.,** *J. Chem. Inf. Comput. Sci.,* 20, 106, 1980.

Chapter 8

SUBSPECTRAL MOLECULES

An examination of tabulated graph spectra[1-3] reveals that the occurrences of isospectral molecular graphs are rare (for small N). A much more common situation is that in which the spectra of two different molecules have one or more common eigenvalues. In some cases, the spectrum of one molecular graph contains the complete spectrum of a second, smaller (component) graph. The two graphs are then said to be *subspectral*.[4] Several authors, in discussing the eigenvalues of the Hückel matrix, encountered the isospectrality but did not recognize this regularity as the connection between the Hückel graph (molecule) and its constituting subgraphs (fragments).[5,6] Hall[7] even identified subspectrality with isospectrality. This is in a way so, because in the subspectral pair, the spectrum (or a part of it) of the smaller graph is the part of the spectrum of greater graph. But, in order to avoid confusion with isospectral graphs for this case we use the term *subspectrality*.[8]

The subspectral relationship does not require that two molecules have the same number of vertices or edges or the same number of rings. The most important examples for chemistry would be those in which the highest occupied molecular orbitals (HOMOs) are the same in two molecules and the lowest unoccupied molecular orbitals (LUMOs) also match each other. Because of the pairing of bonding and antibonding energy levels in alternant hydrocarbons, the coincidence of the LUMO is guaranteed if the HOMO levels match. A survey of the compilation of Hückel MO calculations reveals many examples.[1-3]

Consider the eigenvalues of the bonding orbitals of butadiene, metadivinylbenzene, naphthalene, and [10]-annulene (**1** to **4**).

1

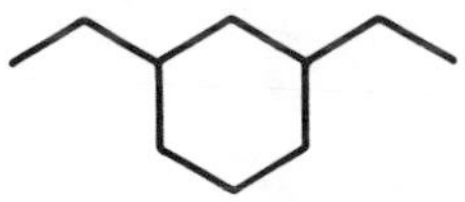

2

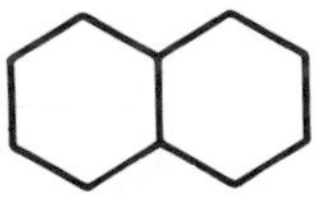

3

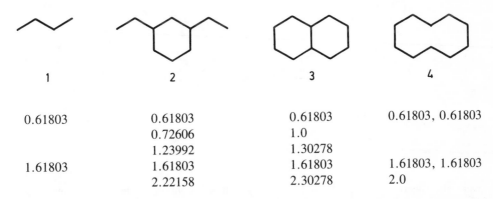

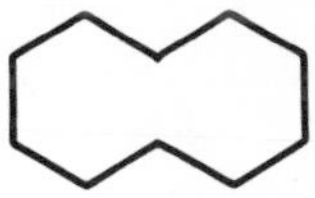

4

1	2	3	4
0.61803	0.61803	0.61803	0.61803, 0.61803
	0.72606	1.0	
	1.23992	1.30278	
1.61803	1.61803	1.61803	1.61803, 1.61803
	2.22158	2.30278	2.0

Structures **2** to **4** have an equal number of vertices and an equal number of eigenvalues, but they are not isospectral. They are, however, *partially isospectral* because their spectra contain the eigenvalues of butadiene, with which each is subspectral. In many (but not all) cases, we can identify the smaller molecule as a fragment of the larger one, just as **1** is a fragment in **2** to **4**, with all the eigenvalues of the fragment appearing in the spectrum of the larger molecule. The trivial example of partially isospectral molecules are those molecules whose spectra share only the zero eigenvalues of the nonbonding molecular orbitals.

Chemically interesting cases of subspectrality are those in which both the fragment and the larger composite molecule are nonradicals and in which neither has nonbonding MOs. Therefore, we will consider only those examples in which both fragment and composite molecule have an even number of vertices.

There are several general schemes by which the spectrum of the fragment can be shown to be a part of that of the larger molecule.

I. LINEAR AND CYCLIC POLYENES

The energy levels of the linear system of N vertices appear in the spectrum of the cyclic system containing 2N + 2 vertices. Compare, for example, structures **1** and **4**. The energy levels of linear and cyclic polyenes are known in analytical form, polyenes

i. *Linear polyenes*

$$x_j = 2\cos\left(\frac{j\pi}{N+1}\right) ; j = 1,2,\ldots,N \tag{1}$$

ii. *Cyclic polyenes*

$$x_j = 2\cos\left(\frac{2j\pi}{N}\right) ; j = 1,2,\ldots,N \tag{2}$$

Suppose we connect two linear polyenes of degree N with two additional vertices plus adjoining edges to form a cyclic polyene of the size L = 2N + 2,

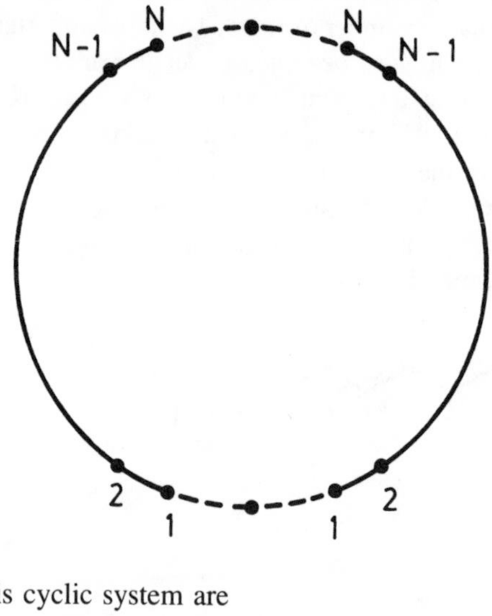

The energy levels of this cyclic system are

$$x_j = 2\cos\left(\frac{2j\pi}{2N+2}\right) = 2\cos\left(\frac{j\pi}{N+1}\right) ; j = 0,1,\ldots,2N+2 \tag{3}$$

Thus, the spectrum of the 2N + 2 cyclic system will contain all the eigenvalues of the N linear system.

Note, that from these arguments follows the mnemonic algorithm of Frost and Musulin.[5]

II. SYMMETRIC SINGLY BRIDGED FRAGMENTS

Živković and co-workers[4] have used the Heilbronner factorization procedure[9] to show why the eigenvalues of a fragment in some cases appear in the spectrum of the larger composite molecule. Consider the composite molecule G formed from two identical fragments A linked through a single vertex,

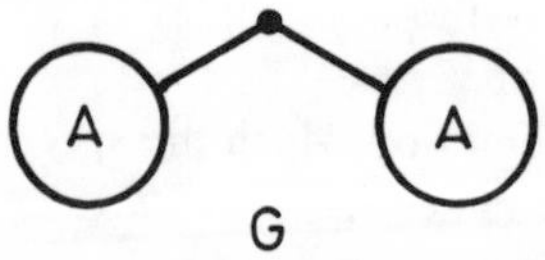

Using Equation (1) of Chapter 7, the characteristic polynomial of G can be written as,

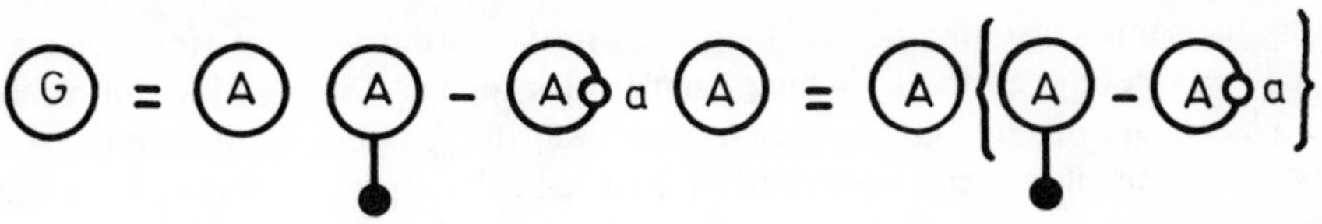

(4)

The characteristic polynomial of fragment A factors out of that for G and therefore the eigenvalues of A will appear in the spectrum of G. This example is not very interesting because G has an odd number of vertices and therefore a nonbonding MO or a zero in its spectrum. The situation can be remedied by introducing the modification G′ in which the substitutent B contains an odd number of vertices and A has an even number,

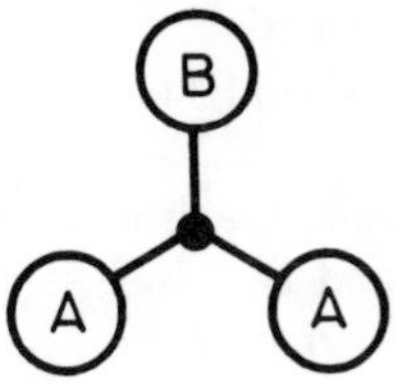

G′

Then, the characteristic polynomial of G′ is given by,

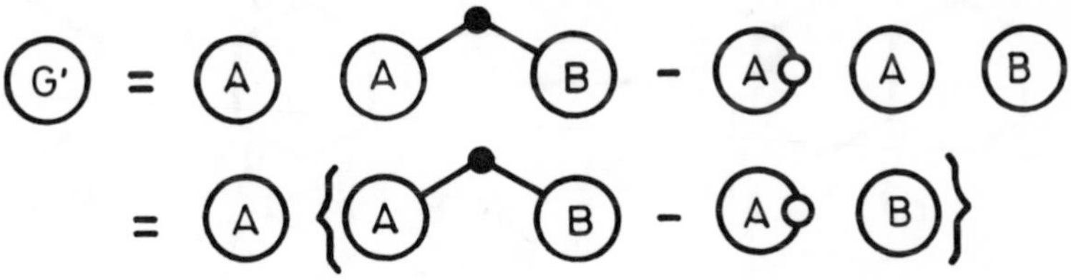

(5)

The polynomial of A factors out of that for G′, and, therefore, the spectrum of A will be contained by the spectrum of G′. An example is structure **5,** in which the fragment A is benzene and B is a single vertex.

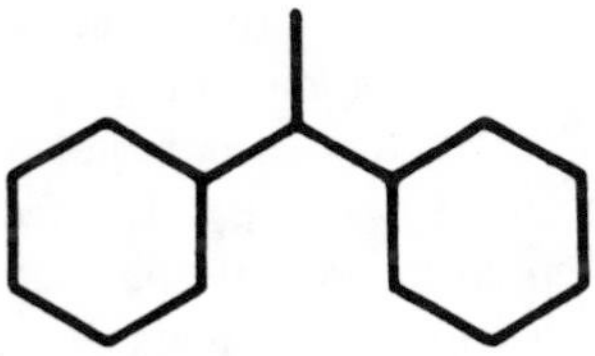

5

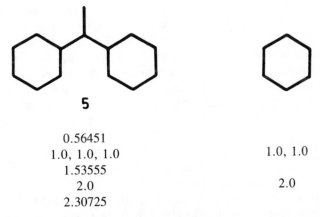

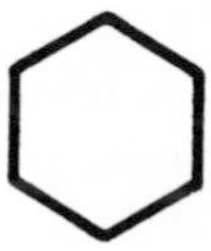

0.56451	
1.0, 1.0, 1.0	1.0, 1.0
1.53555	
2.0	2.0
2.30725	

Although the benzene spectrum is a component of that of **5,** the HOMOs of the two molecules do not match.

III. FRAGMENTS LINKED BY MULTIPLE BRIDGES

Using the same reasoning as that employed in the earlier work of Heilbronner,[9] McClelland[10] has developed the following simple rules for the decomposition of a molecule with a plane of symmetry into simpler fragments for the purpose of simplifying the characteristic polynomial or secular determinant. McClelland's approach allows one to draw two component graphs by fragmenting the composite graph along its plane of symmetry. The rules for the fragmentation process may be stated as follows:

1. The symmetry plane perpendicular to the plane of the graph divides the composite graph into two fragments or component graphs, A and B.
2. Vertices lying on the plane of symmetry are included in the A fragment.
3. The weight of the edge[11] between a vertex lying on the plane of symmetry and vertex not on the plane is $\sqrt{2}$.
4. If the symmetry plane bisects an edge between vertices μ and μ', then vertex μ (in A) is weighted $+1$ and vertex μ' (in B) is weighted[11-13] -1. All other edge and vertex weights remain unchanged.
5. The eigenvalues of the composite graph may be obtained by finding the eigenvalues of the component graphs A and B, which, should symmetry allow, might also be factorable by the application of rules 1 through 4.

McClelland's procedure may be illustrated by the application of these rules to the naphthalene (**3**) and cyclopentadienyl (**6**) graphs, as shown below.

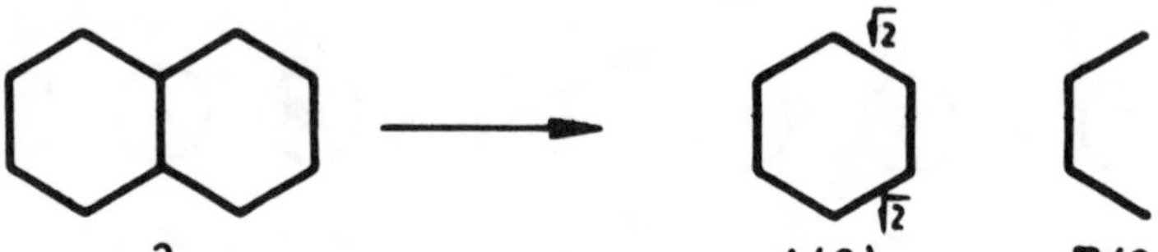

3 A(3) B(3)

6 A(6) B(6)

In the diagram above, X indicates an atom of weight $+1$ (or in the language of Hückel theory coulomb integral of $\alpha + \beta$) while 0 denotes an atom of weight -1 (or coulomb integral of $\alpha - \beta$). The quantity $\sqrt{2}$ written next to a bond indicates that the bond has a weight of $\sqrt{2}$ (the resonance integral for the bond is $\sqrt{2}\ \beta$). The factorization scheme described above is derived from a standard group theoretical approach to HMO theory. McClelland demonstrates that the character table for the C_2 point group may be used to generate, from the N $2p_z$ atomic orbitals $\Theta_1, \Theta_2, \ldots, \Theta_N$ comprising the molecular skeleton, N symmetry-adapted orbitals $\Phi_1, \Phi_2, \ldots, \Phi_N$, which belong to either the A or B representation.

Let atomic orbitals paired by the symmetry operation be denoted by Θ_μ and $\Theta_{\mu'}$, and let the orbitals self-equivalent under the operation (that is, the orbitals lying on the symmetry plane) be denoted by Θ_k. The symmetry orbitals belonging to the A-representation then have the form Θ_k or $(1/\sqrt{2})(\Theta_\mu + \Theta_{\mu'})$; the symmetry orbitals belonging to the B-representation have the form $(1/\sqrt{2})(\Theta_\mu - \Theta_{\mu'})$. Notice that, if atoms μ and ν are adjacent, then so are μ' and ν'; furthermore, $\langle\Theta_\mu|\hat{H}|\Theta_\nu\rangle = 0$ unless μ and ν represent the same atom or adjacent atoms. From these facts it follows that the matrix components of the Hamiltonian on the symmetry-adapted basis are given by: $<\Phi_\mu|\hat{H}|\Phi_\nu> = <\Theta_\mu|\hat{H}|\Theta_\nu> = \beta$ if both or neither of the adjacent atoms μ and ν lie on the plane of symmetry; $<\Phi_\mu|\hat{H}|\Phi_\nu> = <\Theta_k|\hat{H}|(1/\sqrt{2})(\Theta_\nu + \Theta_{\nu'}) = \sqrt{2}\beta$ if one of a pair of adjacent atoms lies on the plane of symmetry; and

$$<\Phi_\mu|\hat{H}|\Phi_\mu> = <\Theta_\mu|\hat{H}|\Theta_\mu> \pm <\Theta_\mu|\hat{H}|\Theta_{\mu'}> = \begin{cases} \alpha \text{ if } \mu \text{ and } \mu' \text{ are nonadjacent} \\ \alpha \pm \beta \text{ if } \mu \text{ and } \mu' \text{ are adjacent} \end{cases} \quad (6)$$

Given the relationship between the Hückel parameters and the elements of the adjacency matrix of an edge- and vertex-weighted graph,[11-13] the origin of McClelland's factorization rules is apparent.

Let us now consider fragments linked by multiple bridges. Consider the molecule M formed from a pair of identical fragments R linked across a plane of symmetry by *n* vertices,

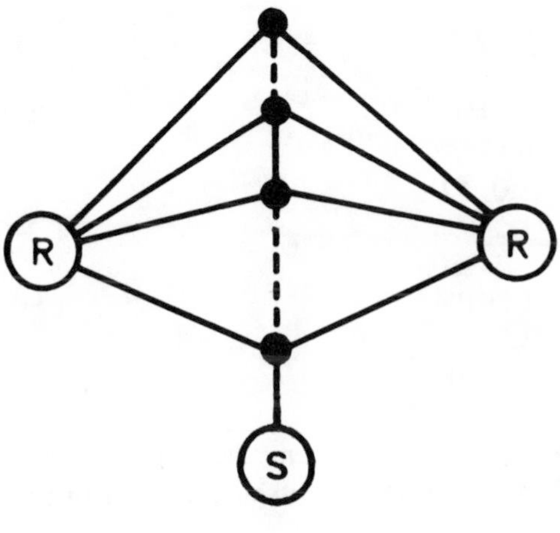

M

The bridging vertices may be bonded pairwise or they may have attached substituents S which if present must be symmetrical with respect to the symmetry plane. Now divide M into two fragments A and B subject to the rules (2) and (3) above.

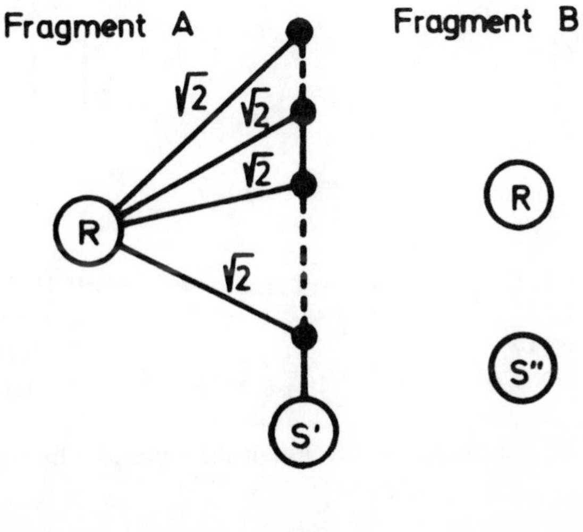

The eigenvalues of fragments (R) and (S″), assuming S is properly divided will, therefore, appear in the spectrum of the composite molecule M. Structures **7** and **8** serve as an example.

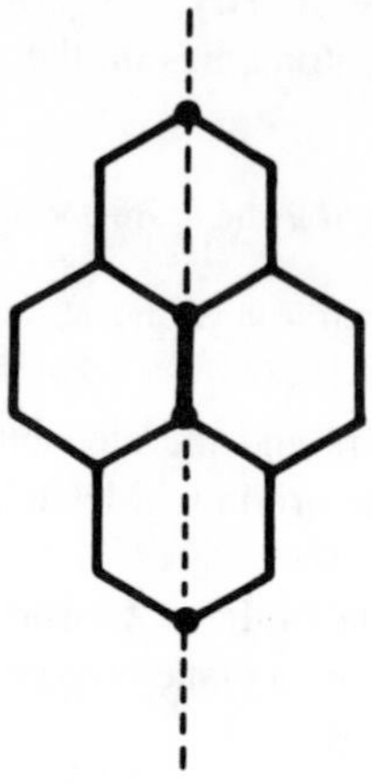

7

8

7	8
0.44504	0.44504
0.87938	
1.0, 1.0	
1.24698	1.24698
1.34730	
1.80194	1.80194
2.0	
2.53209	

Two different composites may share eigenvalues because they share the same fragments Simple examples are **2, 3,** and **4,** but the situation in the pair **9** and **10** is less obvious.

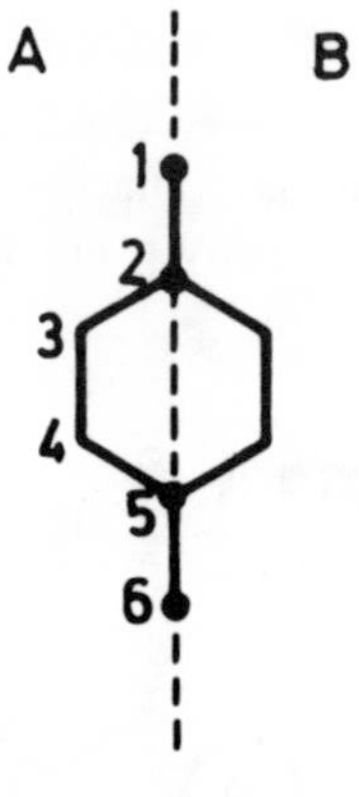

9

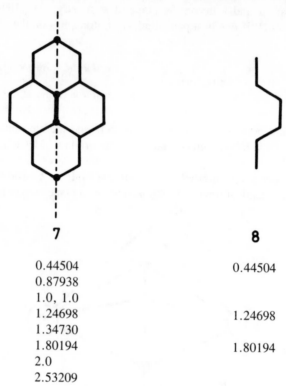

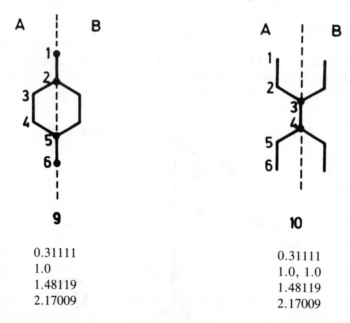

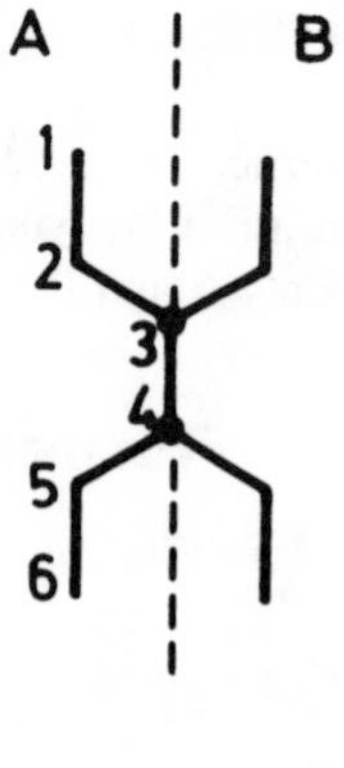

10

9	10
0.31111	0.31111
1.0	1.0, 1.0
1.48119	1.48119
2.17009	2.17009

Structures **9** and **10** are not isospectral because of the repeated ethylene roots in **10**

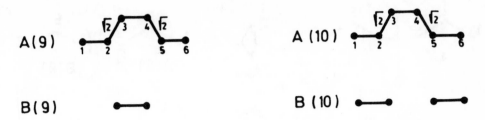

A(9) 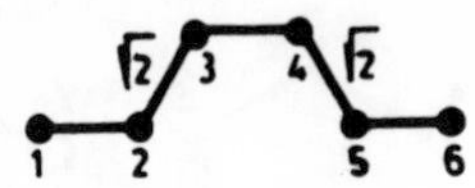

A (10) 

B(9) 

B (10) 

The B fragments of **9** and **10** are, respectively, one and two ethylenes. Therefore, one set of ± 1.0 eigenvalues appears in the spectrum of **9** and two sets in **10.** The A fragments of **9** and **10** are identical. They are related to hexatriene, but with resonance integrals for 2–3 and 4–5 connections increased by the factor $\sqrt{2}$.

The preceding examples all involve molecules in which vertices lie on the symmetry plane and no bonds cross the symmetry plane. These requirements ensure that at least the smaller B fragment will be obtained from the division with coulomb and resonance integrals unchanged from those of the parent molecule.

If the symmetry plane cuts through bonds, then coulomb integrals on both sides of the plane are changed (Rule 4) and the resulting pieces are no longer simple fragments of the composite. Consider **8** and **11**.

8

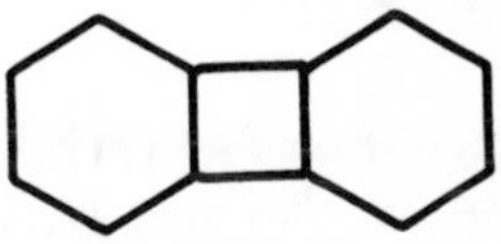

11

8	11
0.44504	0.44504
	0.87938
1.24698	1.24698
	1.34730
1.80194	1.80194
	2.53209

Structure **8** is clearly a fragment of **11** but decomposition of **11** with the horizontal symmetry plane according to Rule 4 does not yield a simple hexatriene fragment.

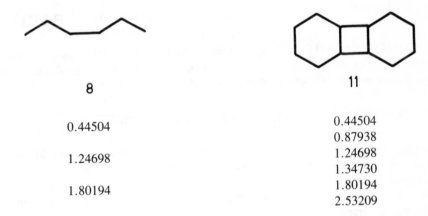

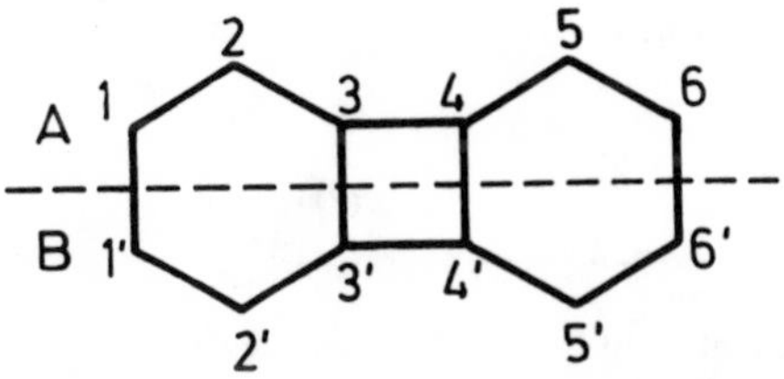

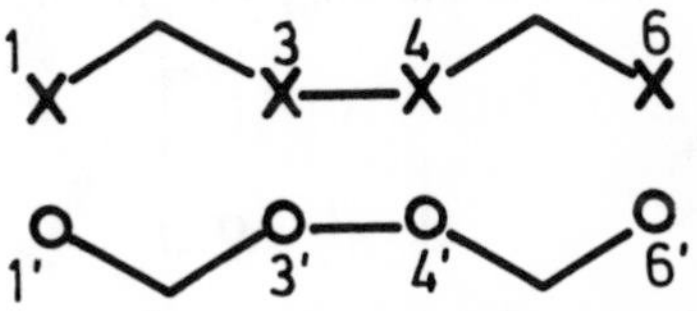

A (11)

B (11)

In the diagram, X signifies an atom with weight +1 (for atoms 1, 3, 4, and 6 of A [**11**]) while 0 denotes −1 weight (as in 1′, 3′, 4′ and 6′ of B [**11**]). Neither A (**11**) nor B (**11**) is identical to **8.**

To see the relationship between the spectra of **8** and **11,** divide **8,** A (**11**) and B (**11**) vertically, cutting the 3–4 bond. Use the symbols Y and P to denote atoms with coulomb integrals $\alpha + 2\beta$ (weight +2) and $\alpha - 2\beta$ (weight −2), respectively.

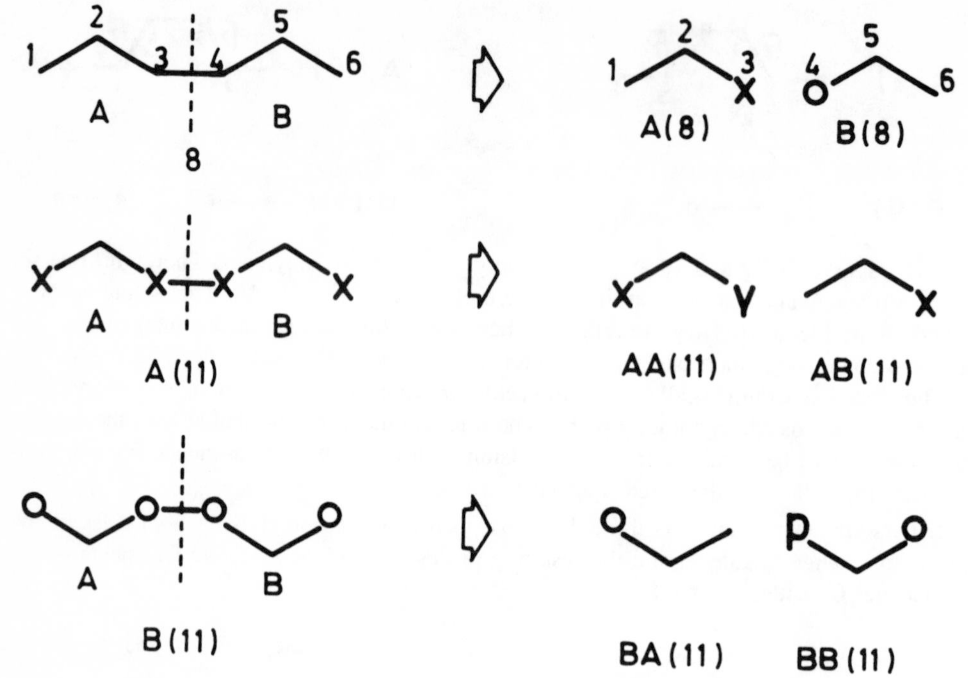

Fragments A (**8**) and AB (**11**) are identical, while B (**8**) matches BA (**11**). Since each member of the two pairs has three vertices, there should be six eigenvalues that coincide between the spectra of **8** and **11,** and exactly that many matchings actually occur.

IV. HALL'S METHOD FOR DETERMINING SUBSPECTRAL GRAPHS

Hall[7] has demonstrated that, in the case of a bipartite graph for which a twofold symmetry operation (i.e., an automorphism p of order 2) exchanges starred and unstarred vertices, one may construct a component graph which is a *contraction* of the composite graph in the sense that its eigenvalues, together with their negatives, constitute the complete spectrum of the composite graph.

The mathematical basis for the construction of such a contraction graph lies in the twofold symmetry properties of the adjacency matrix of the composite graph. Suppose that the 2N vertices of the composite graph are numbered in such a way that the starred vertices are numbered 1 through N and the unstarred vertices are numbered N + 1 through 2N. The adjacency matrix of the composite graph then has the form

$$\mathbf{A} = \begin{bmatrix} \mathbf{O} & \mathbf{B} \\ \mathbf{B}^{T} & \mathbf{O} \end{bmatrix} \tag{7}$$

In order for the twofold symmetry operation, represented by

$$\mathbf{S} = \begin{bmatrix} \mathbf{O} & \mathbf{I} \\ \mathbf{I} & \mathbf{O} \end{bmatrix} \tag{8}$$

to commute with **A**, the submatrix **B** must be symmetrical, that is, $\mathbf{B} = \mathbf{B}^{T}$.

Consider the eigenvalue equation for **A**:

$$\begin{bmatrix} \mathbf{O} & \mathbf{B} \\ \mathbf{B} & \mathbf{O} \end{bmatrix} \begin{bmatrix} u \\ v \end{bmatrix} = x \begin{bmatrix} u \\ v \end{bmatrix} \tag{9}$$

The subvectors u and v represent the coefficients of the starred and unstarred vertices, respectively. Performance of the matrix multiplication gives two equations:

$$\mathbf{B}v = xu \tag{10}$$

$$\mathbf{B}u = xv \tag{11}$$

Since **S** commutes with **A**, $(u\ v)^T$ must also be an eigenvector of **S**; furthermore, the eigenvalues of **S** are $+1$ and -1, since $\mathbf{S}^2 = \mathrm{I}$. Thus the eigenvalue equation for **S**:

$$\begin{bmatrix} \mathbf{O} & \mathbf{I} \\ \mathbf{I} & \mathbf{O} \end{bmatrix} \begin{bmatrix} u \\ v \end{bmatrix} = \pm \begin{bmatrix} u \\ v \end{bmatrix} \tag{12}$$

implies that $u = \pm v$ Substitution of this relation into Equations (10) and (11) produces a new eigenvalue equation:

$$\mathbf{B}u = \pm xu \tag{13}$$

It is clear from Equation (13) that $u = +v$ is an eigenvector of **B** with eigenvalue $+x$ and that $u = -v$ is an eigenvector of **B** with eigenvalue $-x$. Hall concludes from this result that the graph containing N vertices whose adjacency matrix is given by **B** has eigenvectors identical to the u eigenvectors of the original graph (to within a normalization constant) and eigenvalues which, together with their negatives, give the spectrum of the original graph. This smaller graph Hall designates as the *contraction* of the original graph.

Although Hall has not explicitly described the procedure by which a contraction graph may be constructed by mere inspection of a twofold-symmetric composite graph, construction rules may easily be derived upon consideration of the form of the submatrix **B.**

Let p be the mentioned automorphism (the twofold symmetry operation) of the graph G. The contraction graph G′ is defined as follows:

$$V(G') = \left\{ \{v, p(v)\} \,\middle|\, v \in V(G) \right\} \tag{14}$$

$$\left(\text{Note, that } \left| V(G') \right| = \frac{|V(G)|}{2} \text{ since for each } v,\ p(v) \neq v \text{ and } p^2(v) = v \right)$$

$$E(G') = \left\{ \left\{ \{v_1, p(v_1)\}, \{v_2, p(v_2)\} \right\} \,\middle|\, v_1 = v_2 \text{ and } \{v_1, p(v_1)\} \in E(G) \text{ or } v_1 \neq v_2 \text{ and } \{v_1, p(v_2)\} \in E(G) \right\} \tag{15}$$

$$\left(\text{Note that } \{v_1, p(v_2)\} \in E(G) \text{ if and only if } \{v_2, p(v_1)\} \in E(G)\right).$$

$|A|$ means the cardinality of a set A.

The construction of a *contraction graph* is straightforward because it consists of identifying each vertex with its image under symmetry operation; the edges are being left as they are because of the identification. To avoid the loops in the graph G′, instead of the loops we can assign to each vertex with a loop a weight of $+1$ (i.e., a vertex with a loop is every vertex v such that $\{v, p(v)\} \in E(G)$).

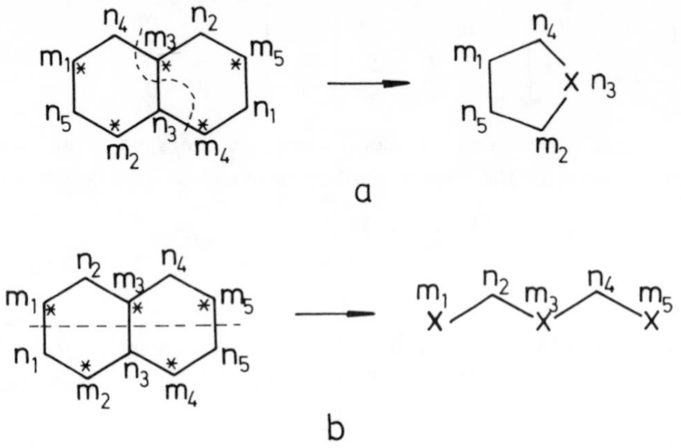

FIGURE 1. Two contractions of the naphthalene graph constructed by Hall's procedure. In (a) the symmetry operation which exchanges starred and unstarred vertices is a twofold rotation and in (b) the twofold operation is reflection through a plane of symmetry.

The algorithm can be illustrated by the construction of two different contraction graphs for the naphthalene graph, as shown in Figure 1. In Figure 1 (a), the symmetry operation which exchanges starred and unstarred vertices is a twofold rotation, whereas in Figure 1 (b), the twofold operation is reflection through a plane of symmetry. Notice that the contraction graph in (b) is identical to the A-fragment of naphthalene as derived by McClelland's procedure. It is clear that Hall's contraction process reduces to McClelland's fragmentation scheme in the case of composite graphs for which the twofold symmetry operation is a plane passing through one or more edges of the graph.

V. FACTORIZATION OF GRAPHS WITH *N*-FOLD SYMMETRY

A third factorization procedure has been developed by D'Amato[14,15] which incorporates both McClelland's and Hall's procedures as special instances of a more general scheme for constructing the subspectral components of composite graphs possessing *n*-fold symmetry. The mathematical origin of this scheme is a unitary transformation upon the eigenvalue equation of the adjacency matrix of the composite graph by the matrix which represents the operation of *n*-fold rotation. King has reported a similar factorization procedure for graphs corresponding to polyhedra.[16] The factorization of twofold- and threefold-symmetric planar composite graphs by this procedure will now be described in detail.

A. Factorization of Graphs with Twofold Symmetry

Let R and S be two sets of vertices such that p (R) = S and vice versa, and for $v \in$ S, $p(v) \neq$ v. The set of vertices v for which $p(v)$ = v will be denoted by Q. The union of sets R, S, and Q is R∪S∪Q = V(G). Let us also suppose that the set of vertices Q creates a discrete graph, i.e., there is no edge between any two vertices of Q. The operation p, which may be rotation about a twofold axis, inversion through a center of symmetry, or reflection through a plane perpendicular to the plane of the graph, need not exchange starred and unstarred vertices if G is a bipartite graph.

The construction of the two graphs G_+ and G_-, which are the factors of G, is as follows. The vertex-sets of G_+ and G_- are the following ones: $V(G_+)$ = R∪Q and $V(G_-)$ = S. If for $v \in$ R is $\{v, p(v)\} \in$ E(G), then v is weighted +1 in G_+, and $p(v)$ is weighted −1 in G_-. Edges with one vertex in Q and the other in R are weighted $\sqrt{2}$ in G_+. Let us consider a pair $\{v_1, v_2\}$, $v_1 \in$ R and $v_2 \in$ R. If $\{v_1, p(v_2)\} \in$ E(G), then the weight is both

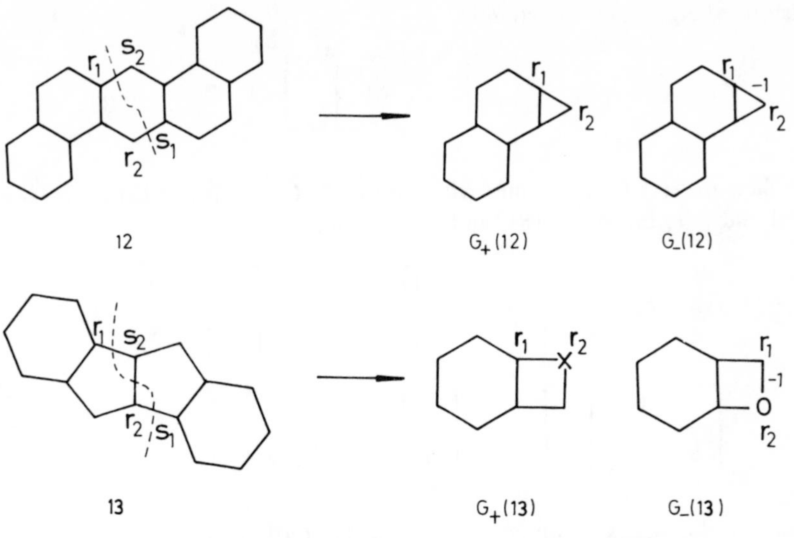

FIGURE 2. Factorization of the twofold-symmetric graphs representing 1,2,5,6-dibenzanthracene (**12**) and 1,2,4,5-dibenzopentalene (**13**) by the procedure of D'Amato.

in G_+ and G_- the same as in G ($+1$ if v_1, v_2 are connected, 0 if they are not). If $\{v_1, p(v_2)\} \in E(G)$, then the weight is increased by 1 in G_+ and decreased by 1 in G_-.

This factorization procedure is applied to the graphs corresponding to 1,2,5,6-dibenzanthracene (**12**) and 1,2,4,5-dibenzopentalene (**13**) in Figure 2. As used earlier, the symbols X and Y refer to atoms with weights $+1$ and $+2$, respectively, while O and P refer to atom weights -1 and -2, respectively.

The procedure for constructing the component graphs by inspection of the composite graph originate in the symmetry properties of the adjacency matrix of the composite graph.

Consider first a twofold-symmetric composite graph which contains 2N vertices, N vertices in each of the equivalent sets R and S. (Exclude, for the present, any graph containing a set of vertices Q which are self-equivalent under the twofold operation.) Let such a graph be denoted by G_1. The vertices of G_1 may be numbered in such a way that the number assigned to each vertex in S is N greater than the number assigned to its symmetry-equivalent partner in R, i.e., $V(G_1) = \{1,2, \ldots ,2N\}$, $1 \leq i \leq N \rightarrow p(i) = i + N$. As a consequence of this numbering procedure, the adjacency matrix of G_1 takes the form:

$$\mathbf{A}(G_1) = \begin{bmatrix} \mathbf{B}_1 & \mathbf{B}_2 \\ \mathbf{B}_2 & \mathbf{B}_1 \end{bmatrix} \tag{16}$$

The elements of the $N \times N$ submatrix $\mathbf{B}_1$ correspond to adjacency relationships within the R set and within the S set; the elements of the $N \times N$ symmetric submatrix $\mathbf{B}_2$ represents the adjacency relationship between sets R and S.

A unitary transformation on $\mathbf{A}(G_1)$ and its eigenvector $(u\ v)^T$ (where the subvectors u and v contain the coefficients of the R and S vertices, respectively) may be performed by the $2N \times 2N$ matrix representing twofold rotation:

$$\mathbf{C}_2 = \begin{bmatrix} \mathbf{0} & \mathbf{I} \\ \mathbf{I} & \mathbf{0} \end{bmatrix} = \mathbf{C}_2{}^{\dagger} \tag{17}$$

The transformed eigenvalue equation is

$$\mathbf{C}_2{}^{\dagger}\,\mathbf{A}(\mathrm{G}_1)\,\mathbf{C}_2\,\mathbf{C}_2{}^{\dagger}\begin{bmatrix} u \\ v \end{bmatrix} = \mathrm{x}\,\mathbf{C}_2{}^{\dagger}\begin{bmatrix} u \\ v \end{bmatrix} \tag{18}$$

From the facts that (1) $\mathbf{C}_2$ commutes with $\mathrm{A(G_1)}$; (2) $\mathbf{C}_2{}^2 = \mathrm{I}$; and (3) the eigenvalues of $\mathbf{C}_2$ are $+1$ and -1, Equation (18) can be rewritten as

$$\mathrm{A}(\mathrm{G}_1)\,\mathbf{C}_2{}^{\dagger}\begin{bmatrix} u \\ v \end{bmatrix} = \mathrm{x}\,\mathbf{C}_2{}^{\dagger}\begin{bmatrix} u \\ v \end{bmatrix} = \pm\,\mathrm{x}\begin{bmatrix} u \\ v \end{bmatrix} \tag{19}$$

or

$$\begin{bmatrix} \mathbf{B}_1 & \mathbf{B}_2 \\ \mathbf{B}_2 & \mathbf{B}_1 \end{bmatrix}\begin{bmatrix} v \\ u \end{bmatrix} = \mathrm{x}\begin{bmatrix} v \\ u \end{bmatrix} = \pm\,\mathrm{x}\begin{bmatrix} u \\ v \end{bmatrix} \tag{20}$$

Performance of the matrix multiplication in Equation (20) gives

$$\mathbf{B}_1\,v + \mathbf{B}_2\,u = \mathrm{x}\,v = \pm\,\mathrm{x}\,u \tag{21}$$

$$\mathbf{B}_2\,v + \mathbf{B}_1\,u = \mathrm{x}\,u = \pm\,\mathrm{x}\,v \tag{22}$$

Clearly, $u = \pm\, v$, and therefore Equations (21) and (22) reduce to a single equation

$$(\mathbf{B}_1 \pm \mathbf{B}_2)\,u = \mathrm{x}\,u \tag{23}$$

or its equivalent

$$(\mathbf{B}_1 \pm \mathbf{B}_2)\,v = \mathrm{x}\,v \tag{24}$$

Thus it can be seen that $u = +v$ is an eigenvector of $\mathbf{B}_1 + \mathbf{B}_2$, and $u = -v$ is an eigenvector of $\mathbf{B}_1 - \mathbf{B}_2$; furthermore, the N eigenvalues of $\mathbf{B}_1 + \mathbf{B}_2$ and the N eigenvalues of $\mathbf{B}_1 - \mathbf{B}_2$ together comprise the spectrum of G_1. The graph G_1 may be factored, then, by constructing the two graphs G_+ and G_-, each containing N vertices, which have as their respective adjacency matrices $\mathbf{B}_1 + \mathbf{B}_2$ and $\mathbf{B}_1 - \mathbf{B}_2$.

The relationship between the forms of $\mathbf{B}_1$ and $\mathbf{B}_2$ and the rules given earlier for constructing the component graphs may best be clarified by means of an example. Consider the 1,2,4,5-dibenzopentalene graph (**13**) shown in Figure 2.

The nonzero elements of $\mathbf{B}_1$ (**13**) correspond to the adjacency relationships between and among the eight vertices within the R set (or within the S set); the nonzero elements of $\mathbf{B}_2$ (**13**) represent the adjacency of vertices r_1 and s_1, r_1 and s_2, and r_2 and s_1.

$$\mathbf{B}_1\ (\mathbf{13}) = \begin{bmatrix} 0&0&0&0&0&0&0&1 \\ 0&0&1&0&0&0&1&0 \\ 0&1&0&1&0&0&0&1 \\ 0&0&1&0&1&0&0&0 \\ 0&0&0&1&0&1&0&0 \\ 0&0&0&0&1&0&1&0 \\ 0&1&0&0&0&1&0&0 \\ 1&0&1&0&0&0&0&0 \end{bmatrix} \tag{25}$$

$$\mathbf{B}_2(13) = \left[\begin{array}{c|cccccc} 1 & 1 & 0 & 0 & 0 & 0 & 0 \\ \hline 1 & & & & & & \\ 0 & & & & & & \\ 0 & & & & & & \\ 0 & & & & \mathbf{0} & & \\ 0 & & & & & & \\ 0 & & & & & & \\ 0 & & & & & & \end{array}\right] \tag{26}$$

It can be seen that the only nonzero diagonal element of $\mathbf{B}_1 \pm \mathbf{B}_2$ will be that corresponding to r_1:

$$(\mathbf{B}_1 \pm \mathbf{B}_2)_{11} = \pm 1 \tag{27}$$

Thus r_1 will be weighted $+1$ in G_+ and -1 in G_-; all other vertices will remain unweighted. All off-diagonal elements of $\mathbf{B}_1 \pm \mathbf{B}_2$ will be identical to the corresponding elements of $\mathbf{B}_1$ except for those elements which represent the edge between r_1 and r_2:

$$(\mathbf{B}_1 \pm \mathbf{B}_2)_{12} = (\mathbf{B}_1)_{12} \pm 1 \tag{28}$$

$$(\mathbf{B}_1 \pm \mathbf{B}_2)_{21} = (\mathbf{B}_1)_{21} \pm 1 \tag{29}$$

Therefore, the weight of the edge between r_1 and r_2 in G_1 (zero in this case) will increase by one unit in G_+ and decrease by one unit in G_-; the weights of all other edges will be unchanged. The origin of the construction (given at the beginning of this section) of the subspectral components of a graph of the form of G_1 should now be apparent.

Suppose the composite graph of interest is identical to a graph of the form of G_1 except that it contains, in addition to the R and S vertices, a set of n vertices (denoted by Q) which are self-equivalent under the twofold operation and not connected among themselves. Let this graph be denoted by G_1'. The adjacency matrix of G_1' can be written in the form,

$$\mathbf{A}(G_1') = \begin{bmatrix} \mathbf{0} & \mathbf{D} & \mathbf{D} \\ \mathbf{D}^T & \mathbf{B}_1 & \mathbf{B}_2 \\ \mathbf{D}^T & \mathbf{B}_2 & \mathbf{B}_1 \end{bmatrix} \tag{30}$$

The nonzero elements of the $n \times N$ submatrices $\mathbf{D}$ represent the edges between the Q vertices and the R and S vertices; the elements of the $N \times N$ submatrices $\mathbf{B}_1$ and $\mathbf{B}_2$ represent the adjacency relationships within and between sets R and S, as previously described. (In the case of molecular graphs, $\mathbf{B}_2$ is generally the zero matrix, since R and S vertices are adjacent only to Q vertices and not to each other.)

The matrix representing twofold rotation must now be written as

$$\mathbf{C}_2' = \begin{bmatrix} \mathbf{I} & \mathbf{0} & \mathbf{0} \\ \mathbf{0} & \mathbf{0} & \mathbf{I} \\ \mathbf{0} & \mathbf{I} & \mathbf{0} \end{bmatrix} = (\mathbf{C}_2')^{\dagger} \tag{31}$$

since rotation about a twofold axis leaves the Q vertices unaffected. The eigenvalue equation for $\mathbf{A}(G_1')$ becomes, under unitary transformation,

$$[(C_2')^\dagger \mathbf{A}(G_1') C_2'] (C_2')^\dagger \begin{bmatrix} p \\ u \\ v \end{bmatrix} = x(C_2')^\dagger \begin{bmatrix} p \\ u \\ v \end{bmatrix} \tag{32}$$

where the subvector p represents the coefficients of the Q vertices. Equation (32) may be written as

$$\begin{bmatrix} \mathbf{0} & \mathbf{D} & \mathbf{D} \\ \mathbf{D}^T & \mathbf{B}_1 & \mathbf{B}_2 \\ \mathbf{D}^T & \mathbf{B}_2 & \mathbf{B}_1 \end{bmatrix} \begin{bmatrix} p \\ v \\ u \end{bmatrix} = x \begin{bmatrix} p \\ v \\ u \end{bmatrix} \tag{33}$$

since $\mathbf{C}_2'$ commutes with $\mathbf{A}(G_1')$ and since $(\mathbf{C}_2')^2 = I$. By performing the matrix multiplication in (33), one obtains Equations (34) through (36):

$$\mathbf{D}\,v + \mathbf{D}\,u = x\,p \tag{34}$$

$$\mathbf{D}^T p + \mathbf{B}_1\, v + \mathbf{B}_2\, u = x\,v \tag{35}$$

$$\mathbf{D}^T p + \mathbf{B}_2\, v + \mathbf{B}_1\, u = x\,u \tag{36}$$

The behavior of the eigenvector under transformation by $(\mathbf{C}_2')^\dagger$ implies that, once again, $u = \pm v$. Therefore, equations (34) through (36) may be simplified in one of two ways, according to whether u is equal to $+v$ or $-v$.

If $u = +v$, then Equations (34) and (36) reduce to

$$2\,\mathbf{D}\,u = x\,p \tag{37}$$

$$\mathbf{D}^T p + (\mathbf{B}_1 + \mathbf{B}_2)\,u = x\,u \tag{38}$$

From Equations (37) and (38), one may reconstruct the matrix equation

$$\begin{bmatrix} \mathbf{0} & 2\,\mathbf{D} \\ \mathbf{D}^T & \mathbf{B}_1 + \mathbf{B}_2 \end{bmatrix} \begin{bmatrix} p \\ u \end{bmatrix} = x \begin{bmatrix} p \\ u \end{bmatrix} \tag{39}$$

which implies that $(N + n)$ of the $(2N + n)$ eigenvalues of G_1' are also eigenvalues of a graph G_+' which has as its adjacency matrix

$$\begin{bmatrix} \mathbf{0} & 2\,\mathbf{D} \\ \mathbf{D}^T & \mathbf{B}_1 + \mathbf{B}_2 \end{bmatrix} \tag{40}$$

It is evident from the form of this matrix that G_+' is identical to the G_+ factor of a graph of the form of G_1 except that the former contains a directed edge of weight 2 from each vertex in Q to its adjacent vertex in R and a directed edge of unit weight from the R vertices to their neighboring Q vertices. Since these directed edges between R and Q vertices do not belong to a cycle in G'_+, only the product of their weights is of interest, and therefore they can be replaced by two directed edges weighted $\sqrt{2}$ each, or, equivalently, by a single undirected edge weighted $\sqrt{2}$.

Referring again to Equations (34) through (36) consider the second alternative, that is, that $u = -v$. In this case, the lefthand side of Equation (34) reduces to zero and the difference between Equations (35) and (36) becomes

$$(\mathbf{B}_1 - \mathbf{B}_2)\,u = x\,u \qquad (41)$$

The second factor of G_1' is the graph whose adjacency matrix is given by $\mathbf{B}_1 - \mathbf{B}_2$; it is therefore identical to the G_- factor of G_1. Construction of the subspectral components of a graph of the form of G_1' follows immediately from this analysis.

It should be noted that, in the case of a bipartite graph for which a twofold symmetry operation exchanges starred and unstarred vertices, the G_+ component constructed by the procedure exposed earlier in this section is identical to Hall's contraction graph; compare, for example, the graph G_+ (**12**), shown in Figure 2, to the contraction graph for 1,2,5,6-dibenzanthracene drawn according to Hall's rules. In the case of graphs for which the symmetry element is a plane perpendicular to the plane of the graph, G_+ and G_- are identical to McClelland's A-fragment and B-fragment, respectively.

The factorization procedure of D'Amato is thus the most general method yet developed for factoring an arbitrary twofold-symmetric composite graph into its subspectral components.

B. Factorization of Graphs with Threefold Symmetry

A procedure for factoring an arbitrary threefold-symmetric composite graph into its subspectral components has been developed by means of mathematical manipulations which are exactly analogous to those employed in the case of twofold-symmetric composite graphs, that is, by performance of a unitary transformation on the eigenvalue equation of the adjacency matrix of the composite graph after numbering the vertices of the composite graph in a manner consistent with its threefold symmetry. The factorization procedure and its derivation will now be described.

Consider a graph G with threefold rotational symmetry operation (automorphism p) and let us split its vertices into 3 sets: R, S, and T ($p[R] = S$, $p[S] = T$, $p[T] = R$), and, possibly, in a set of self-equivalent vertices Q ($p[v] = v$ for $v \in Q$). Two graphs, G_a and G_e, such that the eigenvalues of G_a and eigenvalues of G_e taken twice comprise the complete spectrum of G, can be constructed as follows ($\gamma = \exp[2\pi i/3]$, $\gamma^* = \exp[-2\pi i/3]$): $V(G_e) = R$ and $V(G_a) = R \cup Q$. If $\{v, p(v)\} \in E(G)$, then v is weighted $+2$ in G_a and -1 ($= \gamma + \gamma^*$) in G_e. Edges with one vertex in Q and the other in R are weighted $\sqrt{3}$ in G_a. Let us consider a pair $\{v_1, v_2\}$, $v_1 \in R$ and $v_2 \in R$. If $\{v_1, p(v_2)\} \notin E(G)$, then the weight is (both in G_a and G_e) the same as in G ($+1$ if v_1, v_2 are connected, 0 if they are not). If $\{v_1, p(v_2)\} \in (E(G)$, then the weight is increased by 1 in G_a. In G_e the weight of the directed edge (v_1,v_2) is increased by γ, and the weight of the directed edge (v_2,v_1) is increased by γ^*. Furthermore, if these directed edges do not belong to a cycle in G_e, they may be replaced by an undirected edge whose weight is equal to the square root of the product of the weights of the directed edges, i.e., either $\sqrt{\gamma\cdot\gamma^*} = 1$ or $\sqrt{(1 + \gamma)(1 + \gamma^*)} = 1$. The application of the procedure may be illustrated by the examples given in Figure 3.

The derivation of the procedure will first be given for the type of threefold-symmetric graph which contains 3N vertices, N vertices in each of the equivalent sets R, S, and T (graphs containing a vertex lying on the threefold rotational axis will be discussed in a later section). Let a graph of this type be denoted by G_2. If the vertices of G_2 are numbered in such a way that the number assigned to each vertex in S is N greater than the number assigned to its symmetry partner in R and N less than the number assigned to its symmetry partner in T, i.e., $V(G_2) = \{1,2,\ldots,r.,3N\}$, for $1 \leq i \leq N$, $p(i) = i + N$, and $p^2(i) = i + 2N$, then the adjacency matrix of G_2 has the form,

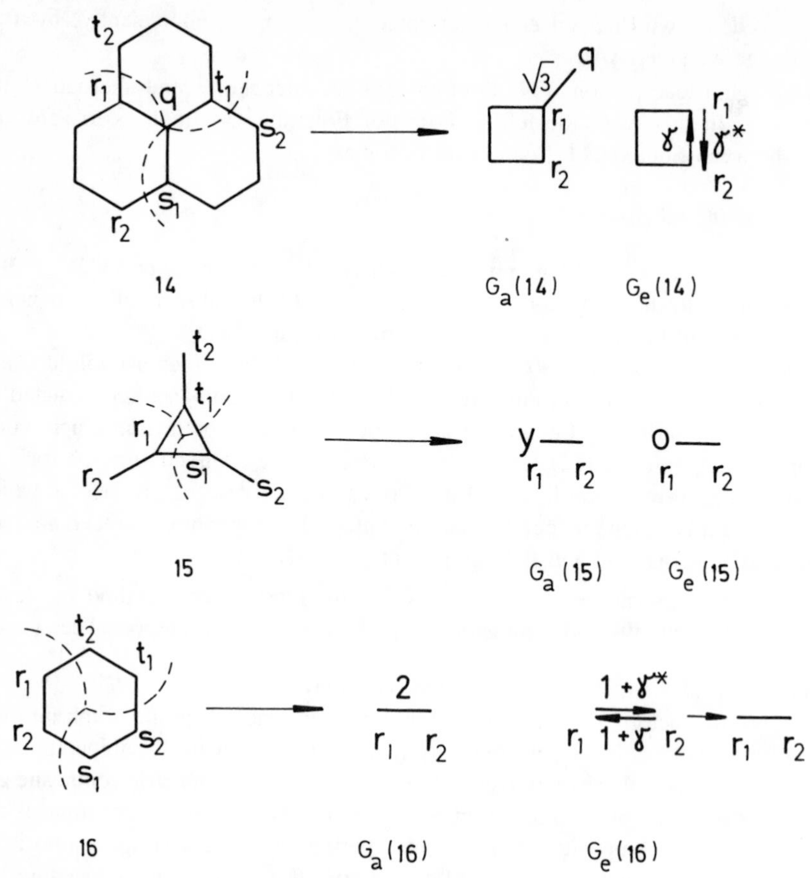

FIGURE 3. Factorization of the threefold-symmetric graphs representing phenalenyl (**14**), trimethylene-cyclopropane (**15**), and benzene (**16**) by the procedure of D'Amato.

$$\mathbf{A}(\mathrm{G}_2) = \begin{bmatrix} \mathbf{B}_1 & \mathbf{B}_2 & \mathbf{B}_3 \\ \mathbf{B}_3 & \mathbf{B}_1 & \mathbf{B}_2 \\ \mathbf{B}_2 & \mathbf{B}_3 & \mathbf{B}_1 \end{bmatrix} \tag{42}$$

The elements of the symmetric submatrix $\mathbf{B}_1$ represent the adjacency relationships within a single set, R, S, or T; the elements of $\mathbf{B}_2$ and $\mathbf{B}_3 = (\mathbf{B}_2)^T$ represent the adjacency relationships between R and S, R and T, and S and T.

The eigenvalue equation for $\mathbf{A}(\mathrm{G}_2)$ is

$$\mathbf{A}(\mathrm{G}_2) \begin{bmatrix} u \\ v \\ w \end{bmatrix} = x \begin{bmatrix} u \\ v \\ w \end{bmatrix} \tag{43}$$

where the subvectors u, v, and w contain the coefficients of the R, S, and T vertices, respectively. A unitary transformation on $\mathbf{A}(\mathrm{G}_2)$ and its eigenvector may be performed by the matrix which represents the operation of threefold rotation:

$$\mathbf{C}_3 = \begin{bmatrix} \mathbf{I} & 0 & 0 \\ 0 & 0 & \gamma\mathbf{I} \\ 0 & \gamma\mathbf{I} & 0 \end{bmatrix} = \mathbf{C}_3^{\dagger} \tag{44}$$

where $\gamma = \exp(2\pi i/3)$.

The transformed eigenvalue equation is

$$[\mathbf{C}_3^{\dagger}\, \mathbf{A}(G_2)\, \mathbf{C}_3]\, \mathbf{C}_3^{\dagger} \begin{bmatrix} u \\ v \\ w \end{bmatrix} = \mathbf{C}_3^{\dagger} \begin{bmatrix} u \\ v \\ w \end{bmatrix} \tag{45}$$

or

$$\begin{bmatrix} \mathbf{B}_1 & \gamma^{*}\mathbf{B} & \gamma\mathbf{B}_2 \\ \gamma\mathbf{B}_2 & \mathbf{B}_1 & \gamma^{*}\mathbf{B}_3 \\ \gamma^{*}\mathbf{B}_3 & \gamma\mathbf{B}_2 & \mathbf{B}_1 \end{bmatrix} \begin{bmatrix} u \\ \gamma w \\ \gamma^{*} v \end{bmatrix} = x \begin{bmatrix} u \\ \gamma w \\ \gamma^{*} v \end{bmatrix} \tag{46}$$

where the identities $\gamma^2 = \gamma^*$, $(\gamma^*)^2 = \gamma$, and $\gamma \cdot \gamma^* = 1$ have been employed. Performance of the matrix multiplication gives three equations:

$$\mathbf{B}_1\, u + (\gamma^{*}\mathbf{B}_3)(\gamma w) + (\gamma\, \mathbf{B}_2)(\gamma^{*} v) = x\, u \tag{47}$$

$$(\gamma\, \mathbf{B}_2) u + \mathbf{B}_1\, (\gamma w) + (\gamma^{*}\mathbf{B}_3)(\gamma^{*} v) = x\, \gamma w \tag{48}$$

$$(\gamma^{*}\mathbf{B}_3) u + (\gamma\, \mathbf{B}_2)(\gamma w) + \mathbf{B}_1\, (\gamma^{*} v) = x\, \gamma^{*} v \tag{49}$$

These equations may be combined in several ways. The linearly independent combinations (47) + γ^*(48) + γ(49), (47) + γ(48) + γ^*(49), and (47) + (48) + (49) lead respectively to equations (50), (51), and (52):

$$(\mathbf{B}_1 + \mathbf{B}_2 + \mathbf{B}_3)\,(u + v + w) = x(u + v + w) \tag{50}$$

$$(\mathbf{B}_1 + \gamma^{*}\mathbf{B}_2 + \gamma\mathbf{B}_3)\,(u + \gamma^{*} w + \gamma v) = x(u + \gamma^{*} w + \gamma v) \tag{51}$$

$$(\mathbf{B}_1 + \gamma\mathbf{B}_2 + \gamma^{*}\mathbf{B}_3)\,(u + \gamma w + \gamma^{*} v) = x(u + \gamma w + \gamma^{*} v) \tag{52}$$

Clearly, the N eigenvalues of the matrix ($\mathbf{B}_1 + \mathbf{B}_2 + \mathbf{B}_3$), the N eigenvalues of ($B_1 + \gamma^* B_2 + \gamma B_3$), and the N eigenvalues of ($\mathbf{B}_1 + \gamma\mathbf{B}_2 + \gamma^*\mathbf{B}_3$) together constitute the complete spectrum of $\mathbf{A}(G_2)$. Furthermore, since equation (52) is the complex conjugate of equation (51) and x is real, the eigenvalues of the matrices ($\mathbf{B}_1 + \gamma^*\mathbf{B}_2 + \gamma\mathbf{B}_3$) and ($\mathbf{B}_1 + \gamma\mathbf{B}_2 + \gamma^*\mathbf{B}_3$) are identical. Thus the graphs which have as their respective adjacency matrices ($\mathbf{B}_1 + \mathbf{B}_2 + \mathbf{B}_3$) = B_a and ($\mathbf{B}_1 + \gamma\mathbf{B}_2 + \gamma^*\mathbf{B}_3$) = $\mathbf{B}_e$ are the subspectral components of G_2.

The relationship between the forms of $\mathbf{B}_a$ and $\mathbf{B}_e$ and the rules for constructing the component graphs $G_{a,2}$ and $G_{e,2}$ may be clarified by means of the following examples. Consider first the benzene graph (**16,** shown in Figure 3) and its adjacency matrix.

$$\mathbf{A(16)} = \begin{bmatrix} 0 & 1 & 0 & 0 & 0 & 1 \\ 1 & 0 & 1 & 0 & 0 & 0 \\ 0 & 1 & 0 & 1 & 0 & 0 \\ 0 & 0 & 1 & 0 & 1 & 0 \\ 0 & 0 & 0 & 1 & 0 & 1 \\ 1 & 0 & 0 & 0 & 1 & 0 \end{bmatrix} = \begin{bmatrix} \mathbf{B}_1 & \mathbf{B}_2 & \mathbf{B}_3 \\ \mathbf{B}_3 & \mathbf{B}_1 & \mathbf{B}_2 \\ \mathbf{B}_2 & \mathbf{B}_3 & \mathbf{B}_1 \end{bmatrix} \tag{53}$$

Notice that $(\mathbf{B}_1)_{12} = (\mathbf{B}_1)_{21} = 1$, since r_1 is adjacent to r_2, and that $(\mathbf{B}_2)_{21} = (\mathbf{B}_3)_{12} = 1$, since r_1 is adjacent to t_2 and r_2 to s_1. $\mathbf{B}_a$ and $\mathbf{B}_e$ thus have the form

$$\mathbf{B}_a = (\mathbf{B}_1 + \mathbf{B}_2 + \mathbf{B}_3) = \begin{bmatrix} 0 & 2 \\ 2 & 0 \end{bmatrix} \tag{54}$$

$$\mathbf{B}_e = (\mathbf{B}_1 + \gamma\mathbf{B}_2 + \gamma^*\mathbf{B}_3) = \begin{bmatrix} 0 & 1+\gamma^* \\ 1+\gamma & 0 \end{bmatrix} \tag{55}$$

These matrices can be considered the adjacency matrices of the edge-weighted graphs G_a(**16**) and G_e(**16**), shown in Figure 3.

As a second example, the graph **15** in Figure 3 has the adjacency matrix,

$$\mathbf{A(15)} = \begin{bmatrix} 0 & 1 & 1 & 0 & 1 & 0 \\ 1 & 0 & 0 & 0 & 0 & 0 \\ 1 & 0 & 0 & 1 & 1 & 0 \\ 0 & 0 & 1 & 0 & 0 & 0 \\ 1 & 0 & 1 & 0 & 0 & 1 \\ 0 & 0 & 0 & 0 & 1 & 0 \end{bmatrix} = \begin{bmatrix} \mathbf{B}_1 & \mathbf{B}_2 & \mathbf{B}_3 \\ \mathbf{B}_3 & \mathbf{B}_1 & \mathbf{B}_2 \\ \mathbf{B}_2 & \mathbf{B}_3 & \mathbf{B}_1 \end{bmatrix} \tag{56}$$

The nonzero elements of $\mathbf{B}_1$ are $(\mathbf{B}_1)_{12} = 1$, since r_1 is connected to r_2; the nonzero elements of $\mathbf{B}_2$ and $\mathbf{B}_3$ are $(\mathbf{B}_2)_{11} = (\mathbf{B}_3)_{11} = 1$, because r_1 is connected to both s_1 and t_1. $\mathbf{B}_a$ and $\mathbf{B}_e$ are therefore given by

$$\mathbf{B}_a = (\mathbf{B}_1 + \mathbf{B}_2 + \mathbf{B}_3) = \begin{bmatrix} 2 & 1 \\ 1 & 0 \end{bmatrix} \tag{57}$$

$$\mathbf{B}_e = (\mathbf{B}_1 + \gamma\mathbf{B}_2 + \gamma^*\mathbf{B}_3) = \begin{bmatrix} \gamma+\gamma^* & 1 \\ 1 & 0 \end{bmatrix} = \begin{bmatrix} -1 & 1 \\ 0 & 0 \end{bmatrix} \tag{58}$$

Again, these may be considered the adjacency matrices of the vertex-weighted graphs G_a (**15**) and G_e (**15**), also shown in Figure 3. The origin of the construction of the component graphs of composite graphs of the form of G_2 should now be apparent.

Suppose that the composite graph of interest is identical to a graph of the form of G_2 except that it contains an additional vertex which lies on the axis of rotation. Let this graph be denoted by G_2'. The adjacency matrix of G_2' can then be written in the form

$$A(G_2) = \begin{bmatrix} 0 & \mathbf{D} & \mathbf{D} & \mathbf{D} \\ \mathbf{D}^T & \mathbf{B}_1 & \mathbf{B}_2 & \mathbf{B}_3 \\ \mathbf{D}^T & \mathbf{B}_3 & \mathbf{B}_1 & \mathbf{B}_2 \\ \mathbf{D}^T & \mathbf{B}_2 & \mathbf{B}_3 & \mathbf{B}_1 \end{bmatrix} \quad (59)$$

The nonzero element of the $1 \times N$ vector **D** represents the edge between q and each of the sets R, S, and T; $N \times N$ submatrices $\mathbf{B}_1$, $\mathbf{B}_2$, and $\mathbf{B}_3$ represent the adjacency relationships within and among the sets R, S, and T, as previously described.

The matrix representing threefold rotation must now be written as

$$C_3' = \begin{bmatrix} 1 & 0 & 0 & 0 \\ 0 & \mathbf{I} & 0 & 0 \\ 0 & 0 & 0 & \gamma\mathbf{I} \\ 0 & 0 & \gamma^*\mathbf{I} & 0 \end{bmatrix} = (C_3')^\dagger \quad (60)$$

since rotation about the threefold axis leaves q unaffected. The eigenvalue equation for $\mathbf{A}(G_2')$,

$$\mathbf{A}(G_2') \begin{bmatrix} p \\ u \\ v \\ w \end{bmatrix} = x \begin{bmatrix} p \\ u \\ v \\ w \end{bmatrix} \quad (61)$$

(where p is the coefficient of the vertex q), becomes under unitary transformation:

$$\begin{bmatrix} 0 & \mathbf{D} & \gamma^*\mathbf{D} & \gamma\mathbf{D} \\ \mathbf{D}^T & \mathbf{B}_1 & \gamma^*\mathbf{B}_3 & \gamma\mathbf{B}_2 \\ \gamma\mathbf{D}^T & \gamma\mathbf{B}_2 & \mathbf{B}_1 & \gamma^*\mathbf{B}_3 \\ \gamma^*\mathbf{D}^T & \gamma^*\mathbf{B}_3 & \gamma\mathbf{B}_2 & \mathbf{B}_1 \end{bmatrix} \begin{bmatrix} p \\ u \\ \gamma w \\ \gamma^* v \end{bmatrix} = x \begin{bmatrix} p \\ u \\ \gamma w \\ \gamma^* v \end{bmatrix} \quad (62)$$

Performance of the multiplication gives

$$\mathbf{D}\, u + (\gamma^*\mathbf{D})(\gamma w) + (\gamma\,\mathbf{D})(\gamma^* v) = x\, p \quad (63)$$

$$\gamma\mathbf{D}^T p + \mathbf{B}_1\, u + (\gamma^*\mathbf{B}_3)(\gamma w) + (\gamma\,\mathbf{B}_2)(\gamma^* v) = x\, u \quad (64)$$

$$\gamma\mathbf{D}^T p + (\gamma\mathbf{B}_2)u + \mathbf{B}_1\,(\gamma w) + (\gamma^*\mathbf{B}_3)(\gamma^* v) = x\,\gamma w \quad (65)$$

$$\gamma^*\mathbf{D}^T p + (\gamma^*\mathbf{B}_3)u + (\gamma\mathbf{B}_2)(\gamma w) + \mathbf{B}_1\,(\gamma^* v) = x\,\gamma^* v \quad (66)$$

Once again, Equations (63) through (66) may be combined in several ways. Consider first a simplified form of Equation (63) and the linear combination (64) + γ^*(65) + γ(66):

$$\mathbf{D}\ (u + v + w) = x\ p \tag{67}$$

$$3\,\mathbf{D}^T\ p + (\mathbf{B}_1 + \mathbf{B}_2 + \mathbf{B}_3)\,(u + v + w) = x\,(u + v + w) \tag{68}$$

From (67) and (68) can be reconstructed the matrix equation:

$$\begin{bmatrix} 0 & \mathbf{D} \\ 3\,\mathbf{D}^T & \mathbf{B}_1 + \mathbf{B}_2 + \mathbf{B}_3 \end{bmatrix} \begin{bmatrix} p \\ u + v + w \end{bmatrix} = x \begin{bmatrix} p \\ u + v + w \end{bmatrix} \tag{69}$$

which implies that (N + 1) of the (3N + 1) eigenvalues of G_2' are also eigenvalues of a graph $G_{a,2}'$ which has as its adjacency matrix the first matrix on the left-hand side of Equation (69). It is evident from the form of this matrix that $G_{a,2}'$ is identical to $G_{a,2}$ except that the former contains a directed edge of unit weight from vertices in R to q and a directed edge of weight 3 from q to vertices in R. Since these directed edges between r and q do not belong to a cycle in $G_{a,2}'$, only the product of their weights is of interest, and therefore they may be replaced by a single undirected edge weighted $\sqrt{3}$.

Consider next a second, linearly independent combination of equations (64) through (66), (64) + (65) + (66):

$$(1 + \gamma + \gamma^*)\mathbf{D}^T\ p + (\mathbf{B}_1 + \gamma\mathbf{B}_2 + \gamma^*\mathbf{B}_3)\,(u + \gamma w + \gamma^* v) =$$

$$x\,(u + \gamma w + \gamma^* v) \tag{70}$$

Since $(\gamma + \gamma^*) = -1$, Equation (70) reduces to Equation (52); the second subspectral component of G_2' has as its adjacency matrix $(\mathbf{B}_1 + \gamma\mathbf{B}_2 + \gamma^*\mathbf{B}_3) = \mathbf{B}_e$ and is therefore identical to $G_{e,2}$. The procedure for constructing the subspectral components of a composite graph of the form of G_2' follow immediately from this analysis.

It should be noted that the mathematical manipulations employed here for the case of graphs with twofold and threefold symmetry can be applied to the treatment of planar graphs with higher-fold rotational symmetry. By numbering the vertices of such graphs in a manner consistent with their symmetry properties, and by choosing an appropriate matrix by which to perform the unitary transformation, rules may easily be developed for factoring symmetric composite graphs into their subspectral components. The general treatment presented here should prove useful for future investigations of graph spectral regularities.

REFERENCES

1. **Streitwieser, A., Jr. and Brauman, J. I.,** *Supplemental Tables of Molecular Orbital Calculations,* Pergamon Press, Elmsford, N.Y., 1965.
2. **Coulson, C. A. and Streitwieser, A., Jr.,** *Dictionary of π-Electron Calculations,* W. H. Freeman, San Francisco, 1965.
3. **Heilbronner, E. and Straub, P. A.,** *Hückel Molecular Orbitals,* Springer-Verlag, Berlin, 1965.
4. **Živković, T., Trinajstić, N., and Randić, M.,** *Croat. Chem. Acta,* 49, 89, 1977.
5. **Frost, A. and Musulin, B.,** *J. Chem. Phys.,* 21, 572, 1953.
6. **Wild, U. P.,** *Theor. Chim. Acta,* 54, 245, 1980.
7. **Hall, G. G.,** *Mol. Phys.,* 33, 551, 1977.
8. **D'Amato, S. S., Gimarc, B. M., Trinajstić, N.,** *Croat. Chim. Acta,* 54, 1, 1981.
9. **Heilbronner, E.,** *Helv. Chim. Acta,* 36, 170, 1953.

10. **McClelland, B. J.,** *J. Chem. Soc. Faraday Trans. 2,* 53, 1974.
11. **Graovac, A., Polansky, O. E., Trinajstić, N., and Tyutyulkov, N.,** *Z. Naturforsch.,* 30a, 1696, 1975.
12. **Rigby, M. J., Mallion, R. B., and Day, A. C.,** *Chem. Phys. Lett.,* 51, 178, 1977; erratum, *Chem. Phys. Lett.,* 53, 418, 1978.
13. **Trinajstić, N.,** *Croat. Chem. Acta,* 49, 593, 1977.
14. **D'Amato, S. S.,** *Mol. Phys.,* 37, 1363, 1979.
15. **D'Amato, S. S.,** *Theor. Chim. Acta,* 53, 319, 1979.
16. **King, R. B.,** *Theor. Chim. Acta,* 44, 223, 1977.

INDEX

A

B

C

D

E

F

G

H